Quantitative
Geography

QUANTITATIVE GEOGRAPHY

Perspectives on Spatial Data Analysis

- A. Stewart Fotheringham
- Chris Brunsdon
- Martin Charlton

SAGE Publications
London • Thousand Oaks • New Delhi

SAGE Publications Ltd
6 Bonhill Street
London EC2A 4PU

SAGE Publications Inc
2455 Teller Road
Thousand Oaks, California 91320

SAGE Publications India Pvt Ltd
32, M-Block Market
Greater Kailash - I
New Delhi 110 048

British Library Cataloguing in Publication data

A catalogue record for this book is
available from the British Library

ISBN 0 7619 5947 5
ISBN 0 7619 5948 3 (pbk)

Library of Congress catalog record available

Typeset by Keytec Typesetting Ltd., Bridport, Dorset
Printed in Great Britain by The Cromwell Press Ltd, Trowbridge, Wiltshire

To our parents

Contents

Preface   xi

1 Establishing the Boundaries   **1**
   1.1 Setting the scene   1
   1.2 What is quantitative geography?   4
   1.3 Applications of quantitative geography   8
   1.4 Recent developments in quantitative geography   10
   1.5 Summary   13
        Notes   14

2 Spatial Data   **15**
   2.1 Introduction   15
   2.2 Spatial data capture   16
   2.3 Spatial objects   17
   2.4 Location on the globe; location on a plane   18
   2.5 Distance   20
   2.6 Representing spatial data   21
   2.7 Models for spatial data   22
        2.7.1 The vector model   22
        2.7.2 The raster model   23
   2.8 Programming with spatial data   24
        2.8.1 Point-in-polygon operations   24
        2.8.2 The use of complex numbers to represent spatial data   25
   2.9 Problems and opportunities   25
   2.10 Summary   29
        Notes   29

3 The Role of Geographical Information Systems   **30**
   3.1 Introduction   30
   3.2 Simple GIS-based spatial analysis   32
        3.2.1 Feature selection by attribute   32
        3.2.2 Feature selection by geometric intersection   33
        3.2.3 Buffering features   34
        3.2.4 Geometric intersection: union   35
        3.2.5 Geometric intersection: intersection   37
        3.2.6 Proximity   38
        3.2.7 Contiguity   41

3.2.8   Interpolation and fields   42
3.2.9   Density functions   45
3.2.10   Analysis on networks   46
3.2.11   Query   49
3.3   Advanced GIS-based spatial analysis   49
3.3.1   Data-integration and management   51
3.3.2   Exploration   54
3.3.3   Post-modelling visualization   55
3.4   Problems   58
3.4.1   Error modelling   58
3.4.2   Cross-area aggregation   59
3.5   Linking higher-order types of spatial analysis to GIS   61
3.6   A role for GIS?   64
Notes   64

4   **Exploring Spatial Data Visually**   **65**
4.1   Introduction   65
4.2   Stem and leaf plots   66
4.3   Boxplots   68
4.4   Histograms   70
4.5   Density estimates   71
4.6   Maps   72
4.7   The scatterplot matrix   75
4.8   Linked plots   77
4.9   Parallel coordinate plots   80
4.10 RADVIZ   82
4.11 Projection pursuit   86
4.12 Summary   91
Notes   92

5   **Local Analysis**   **93**
5.1   Introduction   93
5.2   The nature of local variations in relationships   94
5.3   Measuring local relationships in univariate data   96
5.3.1   Local point pattern analysis   96
5.3.2   Other local measures of univariate spatial relationships   97
5.4   Measuring local relationships in multivariate data   102
5.4.1   Multilevel modelling   103
5.4.2   The spatial expansion method   106
5.4.3   Geographically weighted regression   107
5.5   An empirical comparison of the spatial expansion method and
GWR   114
5.5.1   The data   114
5.5.2   Global regression model results   115
5.5.3   Spatial expansion method results   117
5.5.4   GWR results   118
5.6   Measuring local relationships in spatial interaction models   128
5.7   Summary   129
Notes   129

6   Point Pattern Analysis   **130**
   6.1   Introduction   130
   6.2   Initial exploration   132
         6.2.1   Scatterplots   132
         6.2.2   Other exploratory plots   134
         6.2.3   Non-graphical approaches to point pattern exploration   136
   6.3   Modelling point patterns   138
   6.4   First-order intensity analysis   144
         6.4.1   Kernel density estimates   146
   6.5   Second-order intensity analysis   149
   6.6   Comparing distributions   154
         6.6.1   Comparing kernel densities   155
         6.6.2   Comparing K functions   157
         6.6.3   Comparing a point pattern with a 'population at risk'   159
   6.7   Conclusions   160
         Notes   161

7   Spatial Regression and Geostatistical Models   **162**
   7.1   Introduction   162
   7.2   Autoregressive models   166
         7.2.1   Spatially autoregressive models   167
         7.2.2   Spatial moving average models   169
   7.3   Kriging   171
         7.3.1   The statistical technique   171
         7.3.2   A worked example   175
         7.3.3   Trend surfaces from kriging residuals   176
   7.4   Semi-parametric smoothing approaches   178
   7.5   Conclusions   182

8   Statistical Inference for Spatial Data   **184**
   8.1   Introduction   184
   8.2   Informal inference   185
         8.2.1   Exploratory data analysis (EDA) and visualization   185
         8.2.2   Data mining   187
                 8.2.2.1   Cluster analysis   188
                 8.2.2.2   Neural networks   190
   8.3   Formal inference   193
         8.3.1   Bayesian inference   193
         8.3.2   Classical inference   198
         8.3.3   Experimental and computational inference   201
                 8.3.3.1   Classical statistical inference and spatial
                           autocorrelation   201
                 8.3.3.2   Experimental distributions and spatial
                           autocorrelation   204
                 8.3.3.3   An empirical comparison of classical and
                           experimental inference   206
         8.3.4   Model building and model testing   210
   8.4   Conclusions   211
         Notes   211

9 Spatial Modelling and the Evolution of *Spatial* Theory **213**
  9.1 Introduction 213
  9.2 Spatial interaction as social physics (1860–1970) 215
  9.3 Spatial interaction as statistical mechanics (1970–80) 217
  9.4 Spatial interaction as aspatial information processing (1980–90) 222
  9.5 Spatial interaction as spatial information processing (1990 onwards) 225
  9.6 Summary 234
       Notes 235

10 Challenges in Spatial Data Analysis **236**
  10.1 Retrospect 236
  10.2 Current challenges 237
        10.2.1 The modifiable areal unit problem 237
        10.2.2 Spatial non-stationarity 240
        10.2.3 Alternative inferential frameworks (Bayes, MCMC) 242
        10.2.4 Geometry 243
        10.2.5 Basic spatial concepts: proximity and accessibility 244
        10.2.6 Merging space and time 245
  10.3 Training people to think spatially 246
        10.3.1 Teaching quantitative geography 246
        10.3.2 Software 247
  10.4 Summary 247

Bibliography 249

Index 267

Preface

One of the more puzzling paradoxes that will face those who come to review the development of geography will be why, at the end of the twentieth century, much of geography turned its back on quantitative spatial data analysis just as many other disciplines came to recognize its importance. At a time when geography should have been meeting the rapidly growing demand for spatial data analysts, the majority of its graduates were, at best, non-quantitative and, in quite a few cases, were actively anti-quantitative.

A commonly expressed reason for the negative attitude by many geographers towards one of the discipline's basic elements is a disillusionment on their part with the positivist philosophical underpinnings of much of the early work in quantitative geography. Another, less frequently stated reason, is that spatial data analysis and spatial modelling are perceived to be relatively difficult, not only by students, but also by many academic geographers who typically have non-quantitative backgrounds. Unfortunately, this perception has deterred many researchers from appreciating the nature of the debates which have emerged and which will continue to emerge within modern quantitative geography. This becomes clear in continuing criticisms of quantitative geography which pertain to methodologies that have been surpassed by developments within the field.

This book attempts to redress the rather antiquated view of quantitative geography held by many of those outside the area. Despite what is sometimes perceived from the outside as a relatively static research area, there have in fact been a large number of major intellectual changes within the past decade in quantitative geography. These are often not simply the development of new techniques, which is inevitably happening, but reflect philosophical changes in the way quantitative geography is approached. It is fair to say that some debate has accompanied these changes; one purpose of this book is to describe these developments and to review some of the concomitant issues. In this way, the book portrays quantitative geography as a vibrant and intellectually exciting part of the discipline in which many new developments are taking place and many more await discovery.

This text is, therefore, not intended as a recipe book of quantitative techniques; nor is it meant to be a comprehensive review of *all* of quantitative geography. Rather, it is our aim to provide a statement on the vitality of modern quantitative geography. As such, it provides examples of how quantitative geography, as currently applied, differs from that of twenty, and even ten, years ago. Perhaps the most important role of this book is to provide examples of recent research in quantitative geography where the emphasis has been on the development of techniques explicitly for *spatial* data analysis. What makes the methods of modern quantitative geography different from many of their predecessors is that they have been developed with the recognition that spatial data have unique properties and that these properties make the use of methods borrowed from aspatial disciplines highly questionable. As such, this book acknowledges what the authors see as a turning point in the development of quantitative geography. It is written at a period when quantitative geography has reached a stage of maturity in which its practitioners are no longer primarily importers of other disciplines' techniques but are mainly exporters of novel ideas about the analysis of spatial data.

We hope that by advertising some of the recent developments in spatial analysis and modelling we might foster a greater interest in, and appreciation for, modern quantitative geography. This is particularly the case, for example, when considering developments in visualization, exploratory data analysis, spatial statistical inference and GIS-based forms of spatial analysis.

Inevitably, this book will find its main audience among established quantitative geographers who wish to keep abreast of the rapid developments in the field. However, we hope that it also finds a broader readership, particularly among researchers in related fields who increasingly recognize the need for specialized techniques in spatial data analysis. It should also be useful to non-quantitative geographers who would like to understand some of the current issues and debates in quantitative geography. Perhaps rather ambitiously, we also hope that the book will be read by those geography students who would like to have a better understanding of what quantitative geography can offer as a contribution to making informed decisions related to career paths.

Given the diversity of our intended audience, we recognise that it will be impossible to satisfy every level of readership on every page. Those who are not quantitatively trained are advised to skim some of the more mathematical treatises while those who are quantitatively trained are asked to be tolerant through some of the more descriptive sections.

Finally, the authors would like to express their gratitude to Ann Rooke for her help with some of the figures and to Robert Rojek at Sage Publications for his enthusiasm, encouragement and patience. We are also extremely grateful for the comments of Dave Unwin and Mike Goodchild on earlier drafts of the book. Any remaining errors are, of course, the sole responsibility of the authors.

1 Establishing the Boundaries

1.1 Setting the scene

For many reasons, it is often difficult to write anything definitive about academic trends. Some trends are so short lived that they have relatively little impact; some are cyclical so that their impact at the time of writing is different from that at the time of reading; and some trends exhibit marked variations across countries in both their intensity and their timing so that any comments have limited spatial application. These caveats aside, it is fair to say that quantitative geography generally experienced a 'downturn' in its popularity between the early 1980s and the mid-1990s (Johnston, 1997; Graham, 1997). The reasons for this are difficult to separate and probably include a mix of the following:

1   A disillusionment with the positivist philosophical underpinnings of much of the original research in quantitative geography and the concomitant growth of many new paradigms in human geography, such as Marxism, post-modernism, structuralism and humanism, which have attracted adherents united often in their anti-quantitative sentiments. This disillusionment is very much a phenomenon of human geography: there appears to be no equivalent in physical geography where quantitative methods are generally viewed as an essential component of research. The demise of quantitative human geography has therefore inevitably led to an unfortunate widening of the gap between human and physical geographers because of the lack of any common language or philosophy. As Graf (1998, p. 2) notes:

> While their human geographer colleagues have been engaged in an ongoing debate driven first by Marxism, and then more recently by post-structuralism, post-modernism, and a host of other isms, physical geographers are perplexed, and not sure what all the fuss is about. . . . They do not perceive a need to develop a post-modern climatology, for example, and they suspect . . . that some isms are fundamentally anti-scientific.

2   The seemingly never-ending desire for some new paradigm or, in less polite terms, 'bandwagon' to act as a cornerstone of geographical research. The methodology of quantitative geography, had, for some, run its course by 1980 and it was time to try something new. While it is a strength of geography that the discipline quickly absorbs new trends and research paradigms, it is also a

considerable weakness. The observations of de Leeuw (1994) on social sciences in general are apposite here. In adding to Newton's famous phrase concerning the cumulative nature of research that we stand 'on the shoulders of giants', de Leeuw (p. 13) comments:

> It also means . . . that we stand on top of a lot of miscellaneous stuff put together by thousands of midgets. . . . This is one of the peculiar things about the social sciences. They do not seem to accumulate knowledge, there are very few giants, and every once in a while the midgets destroy the heaps.

3   A line of research that appears to be better accepted in human geography than in some related disciplines is one that is critical of existing paradigms. As quantitative geography was a well-established paradigm, it became, inevitably, a focal point for criticism. Unfortunately, much of this criticism originated from individuals who had little or no understanding of quantitative geography. As Gould (1984, p. 26) notes:

> few of those who reacted against the later mathematical methodologies knew what they were really dealing with, if for no other reason than they had little or no mathematics as a linguistic key to gain entry to a different framework, and no thoughtful experience into the actual employment of such techniques to judge in an informed and reasoned way. Furthermore, by associating mathematics with the devil incarnate, they evinced little desire to comprehend. As a result, they constantly appeared to be *against* something, but could seldom articulate their reasons except in distressingly emotional terms.

4   As part of the broader 'information revolution' which has taken place in society, the growth of geographical information systems (GIS), or what is becoming known as geographical information science (GISc), from the mid-1980s onwards has had some negative impacts on quantitative studies within geography. Interestingly, these negative impacts appear to have resulted from two quite different perceptions of GISc. To some, GISc is seen either as the equivalent of quantitative geography, which it most certainly is not, or as the academic equivalent of a Trojan horse with which quantitative geographers are attempting to reimpose their ideas into the geography curriculum (Johnston, 1997; Taylor and Johnston, 1995). To others, particularly in the USA where geography has long been under threat as an academic discipline, GISc has tended to displace quantitative geography as the paramount area in which students are provided with all-important job-related skills (Miyares and McGlade, 1994; Gober et al., 1995).[1]

5   Quantitative geography is relatively 'difficult' or, perhaps more importantly, is perceived to be relatively difficult both by many academic geographers, who typically have limited quantitative and scientific backgrounds, and by many students. This affects the popularity of quantitative geography in several ways.

It is perceived by many students to be easier to study other types of geography and their exposure to quantitative methodology often extends little beyond a mandatory introductory course. It deters established non-quantitative researchers from understanding the nature of the debates that have emerged and which will continue to emerge within quantitative geography. It also makes it tempting to dismiss the whole field of quantitative geography summarily through criticisms that have limited validity rather than trying to understand it. As Robinson (1998, p. 9) states:

> It can be argued that much of the antipathy towards quantitative methods still rests upon criticisms based on consideration of quantitative work carried out in the 1950s and 1960s rather than upon attempts to examine the more complete range of quantitative work performed during the last two decades.

The relative difficulty of the subject matter might also have encouraged some researchers to 'jump ship' from quantitative geography (for some interesting anecdotes along these lines, see Billinge et al., 1984) as they struggled to keep up with the development of an increasingly wide array of techniques and methods. As Hepple (1998) notes:

> I am inclined to the view that some geographers lost interest in quantitative work when it became too mathematically demanding, and the 'hunter-gatherer' phase of locating the latest option in SPSS or some other package dried up.

This book is written in response to several of the issues raised in the above discussion. Despite being perceived from the outside as a relatively static research area, quantitative geography has witnessed a number of profound changes in the way it is approached. One purpose of this book is to describe not only some of these developments but also the debates surrounding them. In this way, we hope to present a view of quantitative geography as a vibrant, intellectually exciting, area in which many new developments are taking place, and in which many more await discovery.

A second reason for writing the book is that we hope to demonstrate that because of the changes taking place and that have taken place within the subject, several of the well-oiled criticisms traditionally levelled at quantitative geography no longer apply. For instance, the overly simplistic depictions of many that quantitative geographers search for global laws, and that individuals' actions can be modelled without understanding their cognitive and behavioural processes, have rather limited applicability. For those who insist on 'pigeon-holing' everything, modern quantitative geography, with its emphasis on issues such as local relationships, exploratory analysis and individuals' spatial cognitive processes, must be a difficult area to classify.

A third reason is the hope that some of the changes taking place in quantitative geography might make it more appealing to students and by advertising the existence of these developments, we might foster a greater interest in and appreciation for what modern quantitative geography has to offer. This is particularly the case, for example, in the subsequent discussions on topics such as visualization, exploratory data analysis, local forms of analysis, experimental significance testing and GIS-based forms of spatial analysis.

Ultimately, we hope that this book finds readership not just amongst established quantitative geographers who wish to keep abreast of the rapid developments in the field. It should also be useful to quantitative researchers in related disciplines who are increasingly recognizing the need for specialized techniques for handling spatial data. We hope it might also be of some use to non-quantitative geographers who would like to understand some of the current issues and debates in quantitative geography. Finally, it may be of assistance to students who would like to have a better understanding of what quantitative geography can offer, prior to making informed, rather than prejudicial, decisions related to career paths. Given the diversity of our intended audience, we recognize that it will be impossible to satisfy every level of readership on every page. Those who are not quantitatively trained are advised to skim some of the more mathematical sections whilst those who are quantitatively trained are asked to be tolerant through some of the more descriptive sections.

1.2   What is quantitative geography?

Quantitative geography consists of one or more of the following activities: the analysis of numerical spatial data; the development of spatial theory; and the construction and testing of mathematical models of spatial processes. The goal of all these activities is to add to our understanding of spatial processes. This can be done directly, as in the case of spatial choice modelling (Chapter 9) where mathematical models are derived based on theories of how individuals make choices from a set of spatial alternatives. Or, it can be done indirectly, as in the analysis of spatial point patterns (Chapter 6), from which a spatial process might be inferred.

It would perhaps be difficult to claim that the field of quantitative geography is sustained by any deep-rooted philosophical stance or any political agenda. For most of its practitioners, the use of quantitative techniques stems from a simple belief that in many situations, numerical data analysis or quantitative theoretical reasoning provides an efficient and generally reliable means of obtaining knowledge about spatial processes. Whilst it is recognized that various criticisms can be levelled at this approach (and quantitative researchers are often their own sternest critics), it is also recognized that no alternative approach is free of criticism and none comes close to providing the level of information on spatial processes obtained from the quantitative analysis of spatial data. The objective of most studies in quantitative geography is therefore not to produce a flawless piece of research (since in most cases, especially when dealing with social science data, this is impossible), but rather it is to maximize knowledge on spatial processes with the minimum of error. The appropriate question to ask of quantitative research therefore is 'How useful is it?' and not 'Is it completely free of error?'. This does not mean that error is to be ignored. Indeed, the ability to assess error is an important part of many quantitative studies and is obviously a necessary component in determining the utility of an analysis. It does imply, however, that studies can be useful even though they might be subject to criticism.

It might be tempting to label all quantitative geographers as positivists or naturalists (Graham, 1997) but this disguises some important differences in philosophy across the protagonists of quantitative geography. For example, just as some quantitative geographers believe in a 'geography is physics' approach (naturalism) which involves a search for global 'laws' and global relationships, others recognize that there are possibly no such entities. They concentrate on examining variations in relationships over space through what are known as 'local' forms of analysis (Fotheringham, 1998; Fotheringham and Brunsdon, 1999; see also Chapters 5 and 6). This division of belief is perhaps quite strongly correlated with subject matter. Quantitative physical geographers, because their investigations are more likely to involve predictable processes, tend to adopt a naturalist viewpoint more frequently than their human geography counterparts. In human geography, where the subject matter is typically clouded by human idiosyncrasies, measurement problems and uncertainty, the search is not generally for hard evidence that global 'laws' of human behaviour exist. Rather, the emphasis of quantitative analysis in human geography is to accrue sufficient evidence which makes the adoption of a particular line of thought compelling. As Bradley and Schaefer (1998, p. 71) note in discussing differences between social and natural scientists:

> the social scientist is more like Sherlock Holmes, carefully gathering data to investigate unique events over which he had no control. Visions of a positive social science and a 'social physics' are unattainable, because so many social phenomena do not satisfy the assumptions of empirical science. This does not mean that scientific techniques, such as careful observation, measurement, and inference ought to be rejected in the social sciences. Rather, the social scientist must be constantly vigilant about whether the situation being studied can be modeled to fit the assumptions of science without grossly misrepresenting it. ... Thus, the standard of persuasiveness in the social sciences is different from that of the natural sciences. The standard is the compelling explanation that takes all of the data into account and explicitly involves interpretation rather than controlled experiment. The goals of investigation are also different – the creation of such compelling explanations rather than the formation of nomothetic laws.

As well as being less concerned with the search for global laws than some might imagine, quantitative geography is not as sterile as some would argue in terms of understanding and modelling human feelings and psychological processes (Graham, 1997). Current research, for example, in spatial interaction modelling emphasizes the psychological and cognitive processes underlying spatial choice and how we think about space (see Chapter 9). Other research provides information on issues such as the effects of race on shopping patterns (Fotheringham and Trew, 1993) and gender on migration (Atkins and Fotheringham, 1999). There appears to be a strong undercurrent of thought amongst those who are not fully aware of the nuances of current quantitative geography that it is deficient in its treatment of human influences on spatial behaviour and spatial processes. While there is some validity in this view, quantitative geographers increasingly recognize that spatial patterns resulting from human decisions need to account for aspects of human decision-making processes. This is exemplified by the current interest in spatial information processing strategies and the linking of spatial cognition with spatial choice (see Chapter 9). It should also be borne in mind that the actions of humans in aggregate often result from two types of determinants which mitigate against the need to consider every aspect of human behaviour. There are those, such as the deterrence of distance in spatial movement, which can be quantified and applied to groups of similar individuals; and those, such as shopping at a store because of knowing someone who works there, which are highly idiosyncratic and very difficult to quantify. One of the strengths of a quantitative approach is that it enables the measurement of the determinants that can be measured (and in many cases these provide very useful and very practical information for real-world decision making) whilst recognizing that for various reasons, these measurements might be subject to some uncertainty. This recognition of the role of uncertainty is often more important in the applications of quantitative techniques to human geography than to physical geography and makes the former in some ways more challenging and at the same time more receptive to innovative ideas about how to handle this uncertainty.

To some extent the above comments can be made about the use of quantitative methods in other disciplines. What distinguishes quantitative geography from, say, econometrics or quantitative sociology, or, for that matter, physics, engineering or operations research, is its predominant focus on spatial data. Spatial data are those which combine attribute information with locational information (see Chapter 2). For example, a list of soil chemistry properties or unemployment figures is aspatial unless the locations for which the data apply are also given. As described in Chapter 2, spatial data often have special properties and need to be analysed in different ways from aspatial data. Indeed, the focus of this book centres on this very point. Until relatively recently, the complexities of spatial data were often ignored and spatial data were analysed with techniques derived for aspatial data, a classic case of this being regression analysis (see Chapters 5 and 7). What we concentrate on in this book are those areas where techniques and methodologies are being developed *explicitly* for spatial data. Hence, topics such as log–linear modelling and various categorical data approaches, which have been applied to spatial data but which have not been developed with spatial data explicitly in mind, are not covered in this text. The increasing recognition that 'spatial is special' reflects the maturing of quantitative geography from being predominantly a user of other disciplines' techniques to being an exporter of ideas about the analysis of spatial data.

The definition of quantitative geography at the beginning of this section encompasses a great variety of approaches to the subject. Some of these approaches conflict with one another and, where they do, debates arise. We will try to give a flavour of some of these debates in subsequent chapters but the debates have not been particularly acrimonious and generally a *laissez-faire* attitude prevails in which different approaches are more often seen as complementary rather than as contradictory. For example, quantitative geography encompasses both empirical and theoretical research. Advances in theory are typically very difficult to accomplish but are clearly essential for the progression of the subject matter. Obviously, any theoretical development needs to be subject to intense empirical examination, particularly in the social sciences where the general acceptance of theoretical ideas usually takes hold slowly. Typically in geography, as in other disciplines, empirical research has depended on theoretical ideas for its guidance and the dependency is still very much in this direction. However, with the advent of new ideas and techniques in exploratory spatial data analysis (see Chapter 4), empirical research is increasingly being used to guide theoretical development to form a more equal symbiosis. The last decade has probably seen a gradual decline in purely theoretical research in quantitative geography and more of an emphasis on empirical research. To a large part, this change has been brought about by the enormous advances in computational power available to most researchers which has certainly boosted empirical investigations, often computationally very intensive, of large spatial data sets (Fotheringham, 1998; 1999a). However, while computationally intensive methods are revolutionizing some areas of quantitative geography and have made the calibration of theoretical models easier, it has also been argued that in some cases computational power is being relied upon too heavily (Fotheringham, 1998). The 'solutions' to geographical problems found in this way may have limited applicability and may be obtained at the expense of the deeper understanding that comes from theoretical reasoning.

Another division within quantitative geographical research is that between research which is centred on the statistical analysis of spatial data and research focused on mathematical modelling. However, the distinction between what constitutes statistical as opposed to mathematical research can sometimes be blurred and it is perhaps not a particularly important one to make here. A model might, for example, be developed from mathematical principles and then be calibrated by statistical methods. Typically, areas such as the analysis of point patterns (Chapter 6), spatial regression concepts (Chapters 5 and 7) and various descriptive measures of spatial data such as spatial autocorrelation (Chapter 8) are thought of as 'statistical' whereas topics such as spatial interaction modelling (Chapter 9) and location–allocation modelling (Ghosh and Rushton, 1987; Fotheringham et al., 1995) are thought of as 'mathematical'. Statistical methods have come to dominate quantitative geography, particularly in the social sciences, because of the need to account for errors and uncertainty in both data collection and model formulation. Indeed the term 'spatial analysis' is sometimes used as a synonym for quantitative geography although to some the term implies only stochastic forms of analysis rather than deterministic forms of spatial modelling. It is worth noting that a different usage of the term 'spatial analysis' appears to have become commonplace in the GIS field where software systems are advertised as having a suite of data manipulation routines for 'spatial analysis'. However, these routines typically perform geometrical operations such as buffering, point-in-polygon, overlaying and cookie-cutting which form an extremely minor part of what is typically thought of as 'spatial analysis' by quantitative geographers (see Chapter 3).

1.3   Applications of quantitative geography

A major goal of geographical research, whether it be quantitative or qualitative, empirical or theoretical, humanistic or positivist, is to generate knowledge about the processes influencing the spatial patterns, both human and physical, that we observe on the earth's surface. Typically, and particularly so in human geography, acceptance of such knowledge does not come quickly; rather it emerges after a long series of tests to which an idea or a hypothesis is subjected. The advantages of quantitative analysis in this framework are fourfold.

First, quantitative methods allow the reduction of large data sets to a smaller amount of more meaningful information. This is important in analysing the increasingly large spatial data sets obtained from a variety of sources such as satellite imagery, census counts, local government, market research firms and various land surveys. Many spatial data sets can now be obtained very easily over the World Wide Web (e.g. see the plethora of sites supplying spatial data given at *http://www.clark.net/pub/lschank/web/census.html* and in particular the sites of the US Census, *http://www.census.gov*, the US National Imagery and Mapping Agency, *http://www.nima.mil*, and the US Geological Survey, *http://www.usgs.gov*). Summary statistics and a wider body of data reduction techniques (see Chapter 4 for some examples of the latter) are often needed to make sense of these very large, multidimensional data sets.

Secondly, an increasing role for quantitative analysis is in *exploratory data analysis* which consists of a set of techniques to explore data (and also model outputs) in order to suggest hypotheses or to examine the presence of outliers (see Chapter 4). Increasingly we recognize the need to visualize data and trends prior to performing some type of formal analysis. It could be, for example, that there are some errors in the data which only become clear once the data are displayed in some way. It could also be that visualizing the data allows us to check assumptions and to suggest ways in which relationships should be modelled in subsequent stages of the analysis.

Thirdly, quantitative analysis allows us to examine the role of randomness in generating observed spatial patterns of data and to test hypotheses about such patterns. In spatial analysis we typically, although not always, deal with a sample of observations from a larger population and we wish to make some inference about the population from the sample. Statistical analysis will allow such an inference to be made (see also Chapter 8). For instance, suppose we want to investigate the possible linkage between the location of a nuclear power station and nearby incidences of childhood leukaemia. We could use statistical techniques to inform us of the probability that such a spatial cluster of the disease could have arisen by chance. Clearly, if the probability is extremely low then our suspicions of a causal linkage to the nuclear power station are increased. The statistical test would not provide us with a definite answer – we would just have a better basis on which to judge the reliability of our conclusion. Arguably, the use of such techniques provides us with information on spatial patterns and trends in a less tendentious manner than other techniques. For example, leaving inferences to the discretion of an individual after he or she has been presented with rather nebulous evidence is clearly open to a great deal of subjectivity. How the evidence is viewed is likely to vary across individuals. Similarly, the results from quantitative analyses are likely to be more robust than, for example, studies that elicit large amounts of non-quantitative information from a very small number of individuals.

Fourthly, the mathematical modelling of spatial processes is useful in a number of ways. The calibration of spatial models provides information on the determinants of those processes through the estimates of the models' parameters (see Chapters 5 and 9 for examples). They also provide a framework in which predictions can be made of the spatial impacts of various actions such as the building of a new shopping development on traffic patterns or the building of a seawall on coastal erosion. Finally, models can be used normatively to generate expected values under different scenarios against which reality can be compared.

In summary, the quantitative analysis of spatial data provides a robust testing ground for ideas about spatial processes. Particularly in the social sciences, ideas become accepted only very gradually and have to be subject to fairly rigorous critical examination. Quantitative spatial analysis provides the means for strong evidence to be provided either in support of or against these ideas. This is as true in many other disciplines as it is in geography because it is increasingly recognized that most data are spatial in that they refer to attributes in specific locations. Consequently, the special problems and challenges that spatial data pose for quantitative analysis (see Chapters 2 and 10) are increasingly seen as relevant in a variety of subject areas beyond geography (Grunsky and Agterberg, 1992; Goovaerts, 1992; 1999; Cressie, 1993; Krugman, 1996; Anselin and Rey, 1997). Examples include economics, which is increasingly recognizing that many of its applications are spatial; archaeology, where settlement data or the location of artefacts clearly have spatial properties; epidemiology, where space plays an important role in the study of morbidity and mortality rates; political science, where voting patterns often exhibit strong spatial patterns; geology and soil science, where inferences need to be made about data values that lie between sampled points; health care services, where patients' residential locations are important in understanding hospital rationalization decisions; and marketing, where knowledge of the locations of potential customers is vital to understanding store location. For these reasons, quantitative geographers have skills which are much in demand in the real world and are much sought after to provide inputs into informed decision making.

1.4 Recent developments in quantitative geography

The basic reason for writing this book is that quantitative geography has undergone many changes in the last 20 years and particularly in the last decade. These changes have, in some cases, involved fundamental shifts in the way quantitative geographers view the world. However, few outside the area appear to be aware of these changes. Instead they tend to view and to criticize the field for what it was rather than for what it is. While the subsequent chapters will give a much greater feel for the recent changes that have taken place in quantitative geography, this section gives a flavour of some of these discussions.

The development and maturation of GIS has had an effect on quantitative geography, not always in a positive way as noted above and as also commented on by Fotheringham (1999b). In terms of the development of quantitative methods for spatial data, however, the ability to apply such methods within GIS, or at least link the outcome of such methods with GIS, leads to an increase in the potential for gaining new insights (see Chapter 3). As Fotheringham (1999b, p. 23) notes:

> I would argue that it is not necessary to use a GIS to undertake spatial modelling and integrating the two will not necessarily lead to any greater insights into the problem at hand. However, for certain *aspects of the modelling procedure*, integration will have a reasonably high probability of producing insights that would otherwise be missed if the spatial models were not integrated within the GIS.

It is argued that these 'certain aspects of the modelling procedure' for which integration within GIS will be especially beneficial are exploratory techniques (see Chapter 4). Exploratory techniques are used to examine data for accuracy and robustness and to suggest hypotheses which may be tested in a later confirmatory stage. This typical usage can be classified as *pre-modelling exploration*. However, exploratory techniques are not confined to data issues and another use, termed *post-modelling exploration*, is to examine model accuracy and robustness. One relatively simple example of post-modelling exploration with which many readers will already be familiar is the mapping of the residuals from a model in order to provide improved understanding of why the model fails to replicate the data exactly. Clearly, this is a situation where an interactive mapping system would be useful: not only could the map of residuals be viewed but also it could be interrogated. Zones containing interesting residuals could be highlighted to show various attributes of the zone which might be relevant in understanding the performance of the model. Similarly, an aspatial distribution of residuals in one window could be brushed and the brushed values highlighted on a map in a linked window to explore spatial aspects of model performance. Xia and Fotheringham (1993) provide a demonstration of the exploratory use of linked windows in Arc/Info. Further examples of the power of interactive visualization for spatial data are provided in Chapter 4 and by Anselin (1998), Brunsdon and Charlton (1996) and Haslett et al. (1990; 1991).

Current research on visualization in spatial data sets is focused on the need for visualization tools for higher-dimensional spatial data sets (Fotheringham, 1999c). Most visualization techniques have been developed for simple univariate or bivariate data sets (extensions of some of these techniques can be made to visualize trivariate data). However, most spatial data sets have many attributes and hence these relatively simple visualization techniques are inadequate to examine the complexities within such data sets. Relatively few techniques have been developed for more realistic and more frequently encountered hypervariate (having more than three dimensions) data sets (Cleveland, 1993) and the development of such techniques is therefore becoming of greater concern to quantitative geographers. Some examples of visualization techniques for higher-dimensional spatial data sets are provided in Chapter 4.

Another recent and potentially powerful movement within quantitative geography is that in which the focus of attention is on identifying and understanding *differences* across space rather than *similarities.* The movement encompasses the dissection of global statistics into their local constituents; the concentration on local exceptions rather than the search for global regularities; and the production of local or mappable statistics rather than on 'whole-map' values. This trend is important not only because it brings issues of space to the fore in analytical methods, but also because it refutes the rather naive criticism that quantitative geography is unduly concerned with the search for global generalities and 'laws'. Quantitative geographers are increasingly concerned with the development of techniques aimed at the local rather than the global (Anselin, 1995; Unwin, 1996; Fotheringham, 1997a). This shift in emphasis also reflects the increasing availability of large and complex spatial data sets in which local variations in relationships are likely to be more prevalent.

The development of local statistics in geography is based on the idea that when analysing spatial data, it might be incorrect to assume that the results obtained from the whole data set represent the situation in all parts of the study area. Interesting insights might be obtained from investigating spatial variations in the results. Simply reporting one 'average' set of results and ignoring any possible spatial variations in those results is equivalent to reporting a mean value of a spatial distribution without seeing a map of the data. It is therefore surprising that local statistics have not been the subject of much investigation until recently. The importance of the emphasis on 'local' instead of 'global' is presented in Chapter 5 which also includes a detailed description of several examples of locally based spatial analysis. The chapter concentrates on geographically weighted regression, an explicitly spatial technique derived for producing local estimates of regression parameters, which can be used to produce parameter maps from which a 'geography of spatial relationships' can be examined (Brunsdon et al., 1996; 1998a; Fotheringham et al., 1996; 1997a; 1997b; Fotheringham et al., 1998).

Another development in quantitative geography that explicitly recognizes the special problems inherent in spatial data analysis is that of spatial regression models (Ord, 1975; Anselin, 1988). The fact that spatial data typically are positively spatially autocorrelated, that is high values cluster near other high values and low values cluster near other low values, violates an assumption of the classical regression model that the data consist of independent observations. This creates a problem in assessing statistical significance and model calibration: in essence, the errors in the regression model can no longer be assumed to have zero covariance with each other. To counter this problem, Anselin (1988) has suggested two alternative models, a spatial lag model in which the dependent variable exhibits positive spatial autocorrelation and a spatial error model in which the errors in the regression are spatially autocorrelated. These models are described in Chapter 7.

The term 'geocomputation' has been coined to describe techniques, primarily quantitative, within geography that have been developed to take advantage of the recent massive increases in computer power and data (Openshaw and Openshaw, 1997; Openshaw and Abrahart, 1996; Openshaw et al., 1999; Fotheringham, 1998; 1999a; Longley et al., 1998). The term 'computation' carries two definitions. In the broader sense it refers to the use of a computer and therefore any type of analysis, be it quantitative or otherwise, could be described as 'computational' if it were undertaken on a computer. In the narrower and perhaps more prevalent use, computation refers to the act of counting, calculating, reckoning or estimating – all terms that invoke quantitative analysis. The term 'geocomputation' therefore refers to the computer-assisted quantitative analysis of spatial data *in which the computer plays a pivotal role* (Fotheringham, 1998). This definition is meant to exclude fairly routine analyses of spatial data with standard statistical packages (for instance, running a regression program in SAS or SPSS). Under this definition of geocomputational analysis, the use of the computer *drives* the form of analysis undertaken rather than being a convenient vehicle for the application of techniques developed independently of computers. Geocomputational techniques are therefore those that have been developed *with the computer in mind* and which exploit the large increases in computer power that have been, and still are being, achieved.

A simple example, which is developed in Chapter 8 in a discussion of statistical inference, serves to distinguish the two types of computer usage. Consider a spatial autocorrelation coefficient, Moran's I, being calculated for a variable x distributed across n spatial units. Essentially, spatial autocorrelation describes how an attribute is distributed over space – to what extent the value of the attribute in one zone depends on the values of the attribute in neighbouring zones (Cliff and Ord, 1973; 1981; Odland, 1988; Goodchild, 1986). To assess the significance of the autocorrelation coefficient one could apply the standard formula for a t-statistic calculating the standard error of Moran's I from one of two possible theoretical formulae (see Cliff and Ord, 1981; Odland, 1988; Goodchild, 1986; or Chapter 8 for these formulae and examples of their application). Such a procedure is not geocomputational because the computer is simply used to speed up the calculation of a standard error from a theoretical equation. An alternative, geocomputational, technique would be to derive an estimate of the standard error of the autocorrelation coefficient by experimental methods. One such method would be to permute randomly the x variable across the spatial zones and to calculate an autocorrelation coefficient for each permutation. With a sufficiently large number of such auto-correlation coefficients (there is no reason why millions could not be computed but thousands or even hundreds are generally sufficient), an experimental distribution can be produced which allows statistical inferences to be made on the observed autocorrelation coefficient. An example of this type of geocomputational application is given in Chapter 8.

The use of computational power to replace an assumed theoretical distribution has the advantage of avoiding the assumptions underlying the theoretical distribution which may not be met, particularly with spatial data. Consequently, the use of experimental significance testing procedures neutralizes the criticism that hypo-thesis testing in quantitative geography is overly reliant on questionable assump-tions about theoretical distributions. Another criticism of quantitative geography, addressed in Chapter 9, is the assumption that spatial behaviour results from individuals behaving in a rational manner and armed with total knowledge. Perhaps the classic case of this kind of assumption is in spatial interaction modelling where the early forms of what are known as 'gravity models' were taken from a physical analogy to gravitational attraction between two planetary bodies. In Chapter 9 we attempt to show how far we have come since this analogy was made over 100 years ago (although quantitative geography is still criticized for it!). Newer forms of spatial interaction models, based on sub-optimal choices, limited information, spatial cognition and more realistic types of spatial decision-making processes, are described.

1.5 Summary

There are at least two constraints to undertaking quantitative empirical research within geography. One is our limited ability to think about how spatial processes operate and to produce insights that lead to improved forms of spatial models. The other is the restricted set of tools we have to test and refine these models. These tools might be used for data collection (e.g. GPS receivers, weather stations, stream gauges) or for data display and analysis (GIS, computers). In the early stages of computer use, it was relatively easy to derive models that could not be implemented because of the lack of computer power. This was an era when the second constraint was more binding than the first: the level of technology lagged behind our ability to think spatially. We are now no longer in this era. We are now in a situation where the critical constraint is more likely to be our ability to derive new ways of modelling spatial processes and analysing spatial data. The increases in computer power within the last 20 years have been so enormous that the technological constraint is much less binding than it once was. The challenge is now to make full use of the technology to improve our understanding of spatial processes. In many instances the change is so profound that it can alter our whole way of thinking about issues: the development of experimental significance testing procedures and the subsequent decline in the reliance on theoretical distributions is a case in point. The movement from global modelling to local modelling is another.

This book provides a statement on the vitality of modern quantitative geography. It does not, however, attempt to cover all the facets of the subject area. Instead, it concentrates on examples of how quantitative geography differs from the possibly widespread perceptions of it outside the field. In doing so, it provides examples of the research frontier across a broad spectrum of applications where techniques have been developed *explicitly with spatial data in mind*. The book acknowledges a turning point in the development of quantitative geography: it is written at a period when quantitative geographers have matured from being primarily importers of other disciplines' techniques to being primarily exporters of novel ideas and insights about the analysis of spatial data.

Notes

1. The diffusion of both quantitative geography and GISc has been less extensive within the UK where, mainly because of the traditionally more selective nature of university education, geography students have enjoyed relatively good prospects of employment without necessarily having many specific skills. However, this situation is changing very rapidly.

2 Spatial Data

2.1 Introduction

A glance at the shelves of almost any university library will reveal a plethora of books concerned with geographical information systems (e.g. Burrough, 1986; Huxhold, 1991; Laurini and Thompson, 1992; Rhind et al., 1991; Haines-Young et al., 1993; Bonham-Carter, 1994; Martin, 1996; DeMers, 1997; Chrisman, 1997; Heywood et al., 1998).[1] Fundamental to the operation of GIS are spatial data. Although geographers have been using (and abusing) spatial data long before the mid-1980s, there has been a marked diffusion of interest in spatial data handling since then, and an increasing appreciation of the opportunities offered by, and the problems associated with, such data. Given that spatial data are so pervasive, we need to be aware of the nature of spatial data and their interaction with quantitative geography. Indeed, a number of articles in the magazine *GIS Europe* (Gould, 1996) explored briefly the notion that 'spatial is special'. Others also have begun to realize that there are special problems in analysing spatial data (Berry, 1998). This chapter explores some of the issues.

Spatial data comprise observations of some phenomenon that possesses a spatial reference. The spatial reference may be explicit, as in an address or a grid reference, or it may by implicit, as in a pixel in the middle of a satellite image. One form of spatial reference known to almost everyone in the developed world is the address of one's home although few individuals will be able to quote a map reference of their home. However, we normally convert the former into the latter to carry out any processing of such data. This chapter concerns itself first with the nature of spatial data, then with an examination of the opportunities that arise in the analysis of spatial data, and then with a consideration of the problems that confront the would-be spatial data analyst.

Spatial data are not new. Ptolemy was experimenting with spatial data in second-century Egypt when he was attempting to map his world. The early astronomers who were attempting to map the heavens were using spatial data. Attempts at global exploration by various civilizations required knowledge of locations and the means of getting from one place to another. However, the computational facilities at their disposal were rather primitive compared with the desktop computer and the proliferation of software in the last 20 years. The so-called 'GIS Revolution' has led to a more explicit interest in the handling and analysis of spatial data, an interest that has diffused widely outside geography. Geographers may lay first claim to an interest in spatial data, but they have been joined by mathematicians, physicists, geomaticists, biologists, botanists, archaeologists and architects, to name but a few.

Spatial data may mean different things to different users. John Snow's attempt to postulate a particular water pump in Soho (Gilbert, 1958) as the source of contaminated water leading to a cholera outbreak was an early attempt at spatial analysis. He integrated three spatial data sets in a single map: the locations of the streets in Soho, the locations of cholera cases, and the locations of water pumps. It is not known whether he thought of his exercise as 'spatial data analysis', or 'spatial data handling'. Almost every airline's in-flight magazine contains a map of the airline's routes – again, this displays spatial data. Readers of this book will at some time in their life ask another person for an address; here is a different form of spatial data – the address relates to some location on the earth's surface. The UK Post Office postcode system (Raper et al., 1992), originally developed for the automated handling of mail, is another form of spatial data; most people in the UK know their postcode. In the USA, the zip code is used by the US Postal Service in a similar manner. Clearly, spatial data are more common than we might realize yet not all users of spatial data think of their objects of interest as inherently spatial. Anyone who has stepped into a taxi will find in the driver someone who has had to demonstrate a wide familiarity with spatial data (in knowing the locations of streets and landmarks), and a high degree of proficiency in spatial data manipulation (the ability to work out the best route), perhaps far in advance of the capability of any current GIS.

2.2 Spatial data capture

Spatial data arise when we attempt to sample information from the real world. The nature of the sampling is such that we are interested in not only the variation of some phenomenon, but also the location of that variation. We need therefore to sample not just the nature of the phenomenon of interest, but also its location. There is a wide variety of techniques, both manual and automatic, for doing this.

Digitizing is a process that involves the transfer of locational information about features on paper maps into some computer-processable form. This involves a device known as a digitizer, or digitizing tablet. The map is fixed to the surface of the digitizer, and a cursor is moved across the map by hand. The cursor has a pair of cross-hairs in a transparent window which aids precise positioning, and one or more buttons to transfer locational information to a computer. The cross-hairs are positioned above the point whose position is to be digitized, and one of the buttons pressed to send some measurements describing its location on the surface of the digitizer to a computer. Lines are digitized by digitizing points a short distance apart; curves are approximated by a linked series of short line segments. Where a line changes direction, the operator must digitize sufficient points for the changes to be captured with the desired degree of accuracy. Deciding what is 'sufficient' is largely a result of experience. Digitizing is a tedious process if carried out for any length of time. To assist with data capture in this way, many software systems permit subsequent editing of the sampled data to remove errors which may have resulted from the digitizing process. As well as maps, aerial photographs may also be considered as source material for digitizing.

To speed up the data capture process, maps may be scanned, and information from the scanned images obtained using software that will identify points or lines from the map. Whilst small scanners are comparatively inexpensive, scanners for larger maps are costly devices. However, a scanned map or aerial photograph may be used as a backdrop for the display of some other spatial data.

Locational information may also be collected by some method of surveying. Traditionally this involves the use of a theodolite to measure the bearing of one point from another. Modern theodolites permit the storage of data sampled in this way in a small computer attached to the theodolite, from which data can be uploaded into some suitable processing software at the end of the survey. A more recent development has been the global positioning system (GPS), a network of satellites around the earth. A GPS receiver can be used to obtain the location of the point at which the operator is standing by taking bearings from several satellites. Relatively inexpensive GPS receivers tend to be of low accuracy (around 100 m) which may be too poor for the task in hand. However, GPS readings from an accurate receiver can be obtained conveniently and rapidly, and the measurements uploaded to another computer at the end of the survey for further processing.

Another source of spatial data is from 'remotely sensed' imagery. This may be from one of a series of satellites (e.g. LANDSAT, SPOT, AVHRR), or from scanners flown in aircraft. The data are measurements of reflectance of an area on the earth's surface of light of a given wavelength. The area varies in size from a few metres to over a kilometre, depending on the satellite. The resulting images are characterized by large volumes of data (often several tens of megabytes). Remotely sensed imagery usually requires additional processing before it can be conveniently used for analysis. Consideration of the techniques for this is beyond the scope of this volume, and the reader is referred to texts on remote sensing, such as Lillesand and Kiefer (1994), for further information.

If we are interested in the variation of some phenomenon, then we are faced with the problem of how to sample. We usually find that we are sampling either discrete entities (houses, roads, administrative units) or some phenomenon that varies continuously (air pressure, elevation, population density). Conceptually we sometimes refer to the former as objects, and the latter as continuous fields.

2.3   Spatial objects

Simple spatial objects are of three basic types: points, lines or areas. All three are characterized by the existence of some spatial reference describing the location of the entity measured. As well as the spatial reference, there are measurements of the characteristics of the entity being sampled. The decision to use a spatial object of a particular type depends on the nature of what is being observed. If we are interested in modelling the variation in school attainment for some examination at the level of the school, then we may decide to use points as our sampling unit and think of each school as being a point object in our database. Similarly, we may decide to use point objects as the representations of human beings in a database concerned with individual susceptibility to some disease. Line objects may be useful for representing roads or rivers in a spatial database. Sampling the underlying geology of an area may require the use of area objects, where each area represents the spatial extent of some particular geological type.

Measuring a field is a little more difficult. While it is quite easy to identify an individual dwelling or a person, it is less easy to measure population density or air pressure because the variation is spatially continuous. For some simple field data, it may be possible to define a mathematical function that describes the nature of the variation, but in most cases this is either not possible or impracticable. Consequently, fields are often measured in some discretized form such as a set of observations taken at regular intervals on a grid or lattice. Fields may also be measured at irregularly spaced locations.

2.4   Location on the globe; location on a plane

Many of the statistical techniques described in this book require the calculation of distances between objects. This suggests that we should have some consistent means of describing locations on the surface of the earth so that distances can be computed between these locations. We may need other geometrical calculations to provide us with angles or areas.

The conventional means of describing the location of a point on the earth's surface is in terms of the *latitude* and *longitude* of the point. Latitude is measured north from the equator towards the North Pole, and equivalently southward. Locations that share the same latitude are said to lie on the same *parallel* of latitude. Longitude is measured east or west of a line running from the North Pole to the South Pole that passes through Greenwich, England. Locations that share the same longitude are said to lie on the same *meridian*. The meridian that passes through Greenwich (0°) is known as the prime meridian. The units of measurement of latitude and longitude are degrees, minutes and seconds. The computation of distances between places using latitude and longitude is somewhat cumbersome, and spherical trigonometry is required. The distance, s_{ij}, between any two locations on the surface of a sphere is calculated along a great circle using the following formula:

$$s_{ij} = R.\arccos[\cos(90° - \phi_i)\cos(90° - \phi_j)$$
$$+ \sin(90° - \phi_i)\sin(90° - \phi_j)\cos(\lambda_j - \lambda_i)]$$

(2.1)

where R is the radius of the earth, and the latitude and longitude of location i are (ϕ_i, λ_i). The great circle distance is also a geodesic, the shortest distance between pairs of points on a sphere.

It is often more convenient, when analysing spatial data from small parts of the earth's surface, to ignore the curvature of the earth and to consider the data to be lying on a flat plane.[2] Using a Cartesian coordinate system, the distance between two locations may be calculated using Pythagoras' theorem (see Equation (2.4) later), or may be measured manually with a ruler. To convert between a spherical coordinate system and a planar system requires the projection of the coordinates from the sphere onto the plane. There are dozens of different projections, with varying properties and modes of construction (Bugayevskiy and Snyder, 1995; Maling, 1993). Two major classes of projection are conformal and equal area. In a conformal projection, angles between locations are preserved; they are therefore useful for navigation. As the name suggests, the ratios between areas of regions in an equal-area projection are preserved in the spherical and planar representations. This makes them useful for the depiction of political or socio-economic information. Some projections are useful for displaying the globe, others are more useful for displaying a small part of the globe.

Conversion between geographical (latitude/longitude) and planar coordinates in some projection is usually available in GIS software. The conversions for two well-known projections are given below.

1. Mercator (a cylindrical conformal projection):

$$x = R\lambda$$
$$y = R \ln[\tan(\pi/4 + \phi/2)]$$

(2.2)

    where R is the radius of the earth, ln is the logarithm to the base e, ϕ is the latitude of the location, and λ is its longitude.
2. Lambert (a cylindrical equal-area projection):

$$x = R\lambda$$
$$y = R \sin \phi$$

(2.3)

As a convention, latitudes south of the equator and longitudes west of the Greenwich meridian are considered to be negative when calculating projections.

Readers in the UK will be familiar with the maps produced by the Ordnance Survey, the UK's national mapping organization. The projection used for OS maps is a transverse Mercator projection with the central meridian 2° west of the Greenwich meridian, and distances calculated relative to the intersection of the 49°

parallel with the central meridian (the projection's origin). The origin lies off the Brittany/Normandy coast. For convenience, since distances west of the central meridian would be negative, 400 km is added to the x coordinates (which the OS refers to as 'eastings') and 100 km is subtracted from the y coordinates (the 'northings') (Harley, 1975, p. 24). This action gives a false origin for the projection, 40°46'N 7°33'W, a location south-west of the Scilly Isles. The US State Plane Coordinate System is also a transverse Mercator projection; similarly, the US Army Map Service has used a projection known as the Universal Transverse Mercator (UTM) projection for its global mapping. Both projections are based on the Gauss–Krüger projection (Bugayevskiy and Snyder, 1995, p. 163).

There are other ways of projecting the earth's surface. Tobler (1963; 1967) proposed a form of projection known as a cartogram where the criterion is to show features in locations such that the underlying population density is uniform. This has the effect of increasing the extent of areas with high population density relative to those areas with low population density. Dorling (1995) has used cartograms to depict the spatial variation of a wide variety of indicators from the UK 1991 Census of Population. Despite claims that such displays 'show the people', they can be hard to interpret without additional information to help the user locate towns and cities.

2.5 Distance

As mentioned above, a commonly used measure of the distance between two points on a plane is the Euclidean or straight-line distance. If we have two locations whose coordinates are (x_1, y_1), and (x_2, y_2), the Euclidean distance between them is

$$d_{1,2} = \sqrt{(x_1 - x_2)^2 + (y_1 - y_2)^2} \tag{2.4}$$

Following the notation of Gatrell (1983, p. 25), we can rewrite the coordinates of our two locations as (x_{11}, x_{12}), (x_{21}, x_{22}), such that the first subscript refers to the location and the second subscript refers to the coordinate. The Euclidean distance between two locations i and j with coordinates (x_{i1}, x_{i2}) and (x_{j1}, x_{j2}) can now be written as

$$d_E(i, j) = \left[\sum_{k=1}^{2} (x_{ik} - x_{jk})^2 \right]^{1/2} \tag{2.5}$$

In fact, the Euclidean distance can be generalized to m dimensions:

$$d_{\mathrm{E}}(i, j) = \left[\sum_{k=1}^{m}(x_{ik} - x_{jk})^2\right]^{1/2} \tag{2.6}$$

However, the Euclidean distance may not always be the most 'meaningful' measure of spatial separation. To a pedestrian in a city trying to work out the distance between two locations, it fails to take into account the buildings and other obstacles that lie in the path between the two points. To overcome such problems, a family of distance measures exists known as the Minkowski metrics, with the following definition:

$$d_p(i, j) = \left[\sum_{k=1}^{m}|x_{ik} - x_{jk}|^p\right]^{1/p} \tag{2.7}$$

where p is a constant that can have any value from unity to infinity. When $p = 2$, we have the familiar Euclidean distance formula. When $p = 1$ the distance is referred to as the Manhattan, city-block, or taxicab distance. Alternatively, the value of p can easily be estimated from a sample of known road distances. Gatrell (1983, p. 29) presents a table comparing the Euclidean distance, the city-block distance, and the actual route distance for various pairs of French towns. The Euclidean distance is always shorter than the route distance; the city-block distance is greater than the Euclidean distance, but is not always less than the route distance. In this book, distances are Euclidean in two dimensions unless otherwise described.

2.6 Representing spatial data

Data for spatial analysis have two components: one that describes the location of the object of interest, and one that describes the characteristics of the object. We shall refer to these as the spatial component and the attribute component.

A point may be described by an ordered pair of coordinates $\{x, y\}$ defining its location, and an associated vector of attributes **A**. A line may be defined as an ordered set of n points defining its location, $(x_1, y_1; x_2, y_2; x_3, y_3; \ldots, x_n, y_n)$, together with an associated vector of attributes **A**. An area may be defined by a line which describes the location of its boundary, together with a vector of attributes, **A**, of the area itself. Generally, the first and last points of the boundary are coterminous (i.e. $x_1 = x_n$, $y_1 = y_n$).

A field may be defined by a mathematical function indicating the intensity of the phenomenon of interest at any particular location (a, b); thus $z = f(a, b)$. Fields are used in spatial analysis, for example, in the computation of a spatial kernel (Chapters 3 and 5). However, for observations from the 'real world', we usually do not know the form of the mathematical function. A field may be approximated by providing a series of measurements of the phenomenon at regularly spaced intervals (a lattice or grid), or at irregularly spaced intervals. The value of the field at locations where observations have not been made can be obtained by interpolation using techniques commonly referred to as geostatistics (Chapter 7).

2.7 Models for spatial data

There are two commonly encountered models for representing spatial data in a computer, the *vector* model and the *raster* model. We now briefly consider both models.

2.7.1 The vector model

In the vector model, the physical representation of points, lines and areas closely matches the conceptual representation outlined in the previous section. A single point with attributes is represented by an ordered coordinate pair and a vector of attributes. Usually the x coordinate precedes the y coordinate. In some GIS, the geographical coordinates and their associated attributes are kept in separate files, with a unique identifier for each object present in both files to act as a linking mechanism. A line is represented by an ordered set of point coordinates, the first and last of which are sometimes referred to as nodes. In some GIS, lines are referred to as arcs. There will be a vector of attributes for each line. Again, the geographical coordinates and their associated attributes may be in separate files, linked via a unique identifier. Some systems will automatically provide attributes calculated from the geographical data, such as the length of the line, and identifiers for the start and end nodes of the line. An area is an ordered set of point coordinates representing its boundary, with the first and last being identical. As well as the geographical data, there will be an associated set of attributes, some of which may be calculated by GIS (e.g. the area and perimeter of the polygon). There are several different ways of storing area data physically. The simplest is to store the closed polygons, and the associated attribute data, in separate files. This means that the boundaries between adjacent polygons are stored twice. It also means that the determination of adjacency becomes a cumbersome process. A commonly encountered solution is to store internal boundaries only once, with an extra file containing entries that indicate which boundaries should be put together to represent any given polygon in the study area. A by-product of this mode of storage is that determining polygon adjacency is easy (this is required for some of the techniques described in this book).

A variant of line data is network data. A network is a connected set of lines that represents some connected objects such as a river network, or road network, or a rail network. With a river network, the attributes for each line in the network might include the flow along the particular part of the river that it represents, or its

Strahler order number (Strahler, 1952). With the representation of a road network, the attributes might include the class of the road (expressway, highway, lane), the permitted road speed, the time required to travel along that section of road (or two times, representing travel speed in either direction). The first and last vertices in the line would then represent junctions in the road network – we may be interested in whether turning is restricted at a junction. Typical operations on network data include finding the shortest or quickest path between two locations on the network, determining the area within which one may travel from a central point within a given time or distance, or modelling the distribution of goods from a central facility. A common problem, known as the 'travelling salesperson problem', is one in which a route is desired between a series of locations which minimizes either the time or distance to be travelled.

2.7.2 The raster model

An alternative way of storing spatial data is provided by the raster model. Only the data values are stored, ordered either by column within row, or by row within column. This is usually referred to as a lattice or grid. There will be some additional items of data required, in particular the numbers of rows and columns, the coordinates of the origin of the lattice, and the distance between the mesh points. With this information, data from the grid can be related back to the earth's surface. For any given point location, the position of the grid cell containing that point location can be easily calculated, and the information extracted from the grid. No locational information concerning the position of individual grid cells is stored in the raster model.

Representing points, lines and areas in a grid suffers from the constraints of the grid model. First, the accuracy of location is dependent on the size of grid cells. Secondly, only a single attribute can be stored in the grid. With points, the grid cell corresponding to the location of the point is set to the value of the attribute that is required to be stored. If two points hash to the same grid cell, but have different attributes, a rule will be required to decide which one is stored. The grid cells corresponding to the points defining a line can receive the value of the attribute to be stored. Intermediate grid cells between these can be identified using an algorithm such as Bresenham's algorithm (Bresenham, 1965) and filled. Again, where lines with different values for the attribute meet in the same grid cell, a rule will be required to determine which value is placed in the grid cell. Grid cells entirely contained within the boundary of an area can receive the attribute of that area. Grid cells through which the boundary between two or more polygons passes will require a rule to decide from which polygon the attribute is allocated to the cell.

Which is the most appropriate model to use: raster or vector? To some extent, the answer to this question depends on the nature of the data that are at your disposal. If the data are already in lattice form, for example from a satellite image or a digital elevation model, then the raster representation is perhaps to be preferred. If positional accuracy is an overriding concern, vector representation may be preferred. Many GIS provide routines for conversion between raster and vector modes of representation. There is usually some loss of positional accuracy in raster to vector conversion because of the quantization in the raster model. Lines that should be straight will follow the implicit boundaries between adjacent rasters.

2.8   Programming with spatial data

It has been established that spatial data have special properties. As will be demonstrated, these special properties often lead to traditional methods of analysis being inappropriate. This, in turn, often means that specialized software for handling spatial data has to be developed by the user. The analyst is often forced to write a program in a suitable language (FORTRAN, PASCAL, or C, for example), or use one of the recently developed interactive computing environments such as XLisp-Stat (Tierney, 1990), S-Plus (Venables and Ripley, 1997), or R (Ihaka and Gentleman, 1996). Although this is not a text on programming, the following two sections are provided to give a feel for the special issues involved in programming with spatial data (a much more extensive treatment is provided by O'Rourke, 1998).

2.8.1   Point-in-polygon operations

Suppose that we wish to identify whether any points lie within some simple region, such as a square or circle, or a more complex polygonal region. If the simple region is a square or rectangle whose lower left corner is $\{x_{min}, y_{min}\}$, and the upper right corner is $\{x_{max}, y_{max}\}$, the test is whether any point in it satisfies the constraints

$$x_{min} \leqslant x \leqslant x_{max} \quad \text{and} \quad y_{min} \leqslant y \leqslant y_{max} \tag{2.8}$$

If the simple region is a circle, centred on $\{x_{cen}, y_{cen}\}$, and with radius d, the constraint is

$$(x - x_{cen})^2 + (y - y_{cen})^2 \leqslant d^2 \tag{2.9}$$

With a more complex region, defined by a closed polygon, a routine to determine whether a point lies in a plane region is required (sometimes referred to as a point-in-polygon algorithm). Baxter (1976) has some rather terse FORTRAN code to accomplish this; Sedgwick (1990) presents some rather more elegant code. To speed things up, one might test to determine first whether the point lies inside the rectangle which wholly encloses the arbitrary polygon before applying the test. O'Rourke (1998) is another useful source of spatial data processing algorithms, in this case written in the language C.

2.8.2   *The use of complex numbers to represent spatial data*

Implementing spatial data analysis routines can be cumbersome. For example, to determine the Euclidean distance between two points whose coordinates are stored at locations i and j in the arrays x and y, we might code this thus:

```
real x(...), y(...)
```

```
...
```

```
dx = x(i)-x(j)
```

```
dy = y(i)-y(j)
```

```
dsq = dx*dx + dy*dy
```

```
dist = sqrt(dsq)
```

This can be rather more conveniently coded using the complex type thus:

```
complex p(...)
```

```
...
```

```
dist = abs(p(i)-p(j))
```

If the language or environment that is being used for spatial data supports complex numbers then the implementation of many geometrical techniques is simplified. Brunsdon and Charlton (1996) describe both the representation of spatial data using complex numbers and the implementation of a range of geographical data handling functions in the context of exploratory spatial analysis.

2.9   **Problems and opportunities**

Spatial data provide us with a number of special challenges. One of the lessons learned from 30 years of experimentation by geographers and others is that care and thought are needed in analysing spatial data. The plethora of recent texts with titles such as *Interactive Spatial Data Analysis* (Bailey and Gatrell, 1995), *Spatial Analysis and GIS* (Fotheringham and Rogerson, 1994), *Recent Developments in*

Spatial Analysis (Fischer and Getis, 1997) or *Spatial Analysis: Modelling in a GIS Environment* (Longley and Batty, 1996) suggests that geographers are rethinking their approaches to the analysis of spatial data. Furthermore, geographers' interests in spatial data and associated problems have been paralleled by developments in other disciplines, notably statistics and computer science (Cressie, 1993).

The rapid diffusion of GIS during the 1980s has also fostered an interest by non-geographers in geographical problems. For example, were it not for the curiosity of some paediatric oncologists in northern England, then it is debatable whether Openshaw's geographical analysis machine (GAM) would have appeared, or whether Besag and Newell's complementary approach to the same problem would have been developed (Openshaw et al., 1987, 1988; Besag and Newell, 1991; Fotheringham and Zhan, 1996).

However, the analysis of spatial data is not without its problems. Long gone are the days when the aspiring geographical data analyst would pick up a copy of a text with a beguiling title such as *Multivariate Procedures for the Behavioural Sciences* (Cooley and Lohnes, 1962), punch up the FORTRAN codes, and run his or her data through the computer, unaware of the geographical solecisms which were being committed. There has been a recognition that there are problems peculiar to spatial data analysis, such as spatial autocorrelation, identifying spatial outliers, edge effects, the modifiable areal unit problem, and lack of spatial independence, which require special care.

A fundamental assumption with classical statistical inference, for example, is that of independence: in other words, that the observations under scrutiny are unrelated to each other. However, it has been observed that 'everything is related to everything else, but near things are more related than distant things' (Tobler, 1970). As a result, the assumption of independence of the observations is questionable with spatial data. Suddenly, classical tests pose problems. Cressie (1993, p. 25) quotes a remarkable observation made by Stephan over 60 years ago (Stephan, 1934):

> Data of geographic units are tied together, like bunches of grapes, not separate, like balls in a urn. Of course, mere contiguity in time and space does not of itself indicate independence between units in a relevant variable or attribute, but in dealing with social data, we know that by very virtue of their social character, persons, groups, and their characteristics are interrelated and not independent. Sampling error formulas may yet be developed which are applicable to these data, but until then, the older formulas must be used with great caution. Likewise, other statistical measures must be carefully scrutinized when applied to these data.

This contrasts markedly with a piece of advice that appeared in a geographical text just over 40 years later!

> grab some census data for your town, a factor analysis program and a large computer and then run the data through the program ten, twenty, maybe thirty times, trying all the options.
>
> (Goddard and Kirby, 1976)

Cressie (1993, pp. 14–15) demonstrates the problems of ignoring autocorrelation. If we have a sample of n independent observations drawn from a normal distribution of known variance σ^2, then the unbiased estimator of the unknown population mean is

$$\bar{x} = \sum_{i=1}^{n} x_i / n \qquad (2.10)$$

and the 95% confidence interval for the unknown population mean is given by

$$\bar{x} \pm (1.96)\sigma / \sqrt{n} \qquad (2.11)$$

However, if we have a set of data exhibiting spatial autocorrelation ρ, a simple model for the covariance between any pair of observations x_i and x_j might be

$$\text{cov}(x_i, x_j) = \sigma^2 . \rho \qquad (2.12)$$

The variance of the estimator of the mean is given by

$$\text{var}(\bar{x}) = \left(\sum_{i=1}^{n} \sum_{j=1}^{n} \text{cov}(x_i, x_j) \right) \Big/ n^2 \qquad (2.13)$$

Cressie shows that, on substituting (2.12), this becomes

$$\text{var}(\bar{x}) = (\sigma^2/n) \left[1 + 2 \left(\frac{\rho}{1-\rho} \right) \left(1 - \frac{1}{n} \right) - 2 \left(\frac{\rho}{1-\rho} \right)^2 \left(\frac{1 - \rho^{n-1}}{n} \right) \right] \qquad (2.14)$$

Cressie notes that if $n = 10$ and $\rho = 0.26$, then $\text{var}(\bar{x}) = (\sigma^2/n)[1.608]$, which gives a 95% confidence interval of

$$\bar{x} \pm (1.96 \times \sqrt{1.608})\sigma / \sqrt{n} \qquad (2.15)$$

Thus, if we ignore the positive autocorrelation, the calculated confidence interval using the usual formula in (2.11) will be too narrow. If the data exhibit negative autocorrelation, and we ignore it, then the calculated confidence interval from (2.11) will be too wide.

With positively autocorrelated spatial data the sum of the covariances will be positive; conversely, if the data are negatively spatially autocorrelated (neighbouring observations tend to differ), then the sum of the covariances will be negative. If we assume that our observations are independent then we assume zero spatial autocorrelation. If our data are positively spatially autocorrelated, then the standard error of the mean will be greater than if we had assumed independence. Conversely, if our data exhibit negative spatial autocorrelation, the standard error of the mean will be less than if we assume independence.

The table below shows the variation in the bracketed value in (2.15) for different levels of spatial autocorrelation using a sample size of 100. It would appear that even with relatively low positive spatial autocorrelation, the calculated confidence interval using the usual formula will be too narrow.

ρ	0	0.05	0.15	0.25	0.35	0.45	0.55	0.65	0.75	0.85
100	1.960	2.059	2.276	2.550	2.813	3.165	3.609	4.207	5.096	6.669

This has some important consequences for any interpretation we may decide to make with spatial data. For example, if we want to compare means for two groups, we may test whether there is a sufficient difference between the sample means for us to be confident about asserting that there is a difference in the unknown population means. If the data in both groups are positively spatially autocorrelated, then using the variance formula in (2.14) we may decide that the two sample means are not sufficiently different for us to be confident that there is a difference in population means. However, If we ignore the autocorrelation, then we run an increased risk of making the wrong decision – concluding that the population means are different whereas in fact there is not sufficient evidence to support that decision. If the data exhibit negative spatial autocorrelation, and we ignore that autocorrelation, we may conclude incorrectly that there is no difference between the two population means. The measurement of spatial autocorrelation has been espoused as a theme by a number of geographers and statisticians (Cliff and Ord, 1973; 1981; Griffith, 1987). It is also part of the subject matter of Chapter 6.

Another problem arises when we ignore the special nature of spatial data and use data that have been collected from individuals but are aggregated to some system of zones. This problem, now often referred to as the modifiable areal unit problem, was also noticed by sociologists in the late 1940s (Robinson, 1950; Menzel, 1950). Robinson demonstrated that different conclusions could be drawn if one created a correlation based on aggregated spatial data compared with a correlation based on the data for the individuals. Robinson termed this an 'ecological fallacy'. Many spatial data sets are available only in aggregate form (e.g. most of the output from the censuses in the UK, and the USA[3]). We therefore need to be aware that the results of any spatial analysis might vary if the data are aggregated in some other way. This sensitivity has been demonstrated in simple bivariate analysis (Openshaw and Taylor, 1979), in multivariate spatial analysis (Fotheringham and Wong, 1991) and in spatial modelling (Fotheringham et al., 1995). The ability of some GIS to reaggregate data from one set of areal units to another makes the modifiable areal unit problem even more important.

2.10 Summary

The way in which the special properties associated with spatial data have been viewed marks a watershed in the maturity of quantitative geography. Until fairly recently properties such as the modifiable areal unit problem or spatial non-stationarity have been perceived as rather embarrassing *problems* that have weakened quantitative research using spatial data. However, it is now increasingly recognized that not only do the special properties of spatial data impinge on most, if not all, types of analysis, they also represent great *opportunities*. Spatial data analysis and spatial modelling cannot simply be the application of techniques and models developed in aspatial contexts to spatial problems. The task of spatial analysts is to develop new sets of techniques designed specifically with the properties of spatial data in mind. This is where great opportunities arise for significant new developments and it is a sign of the increasing maturity of quantitative geography that we now recognize this and view spatial data in a positive light. The remainder of this book provides a sampling of the opportunities provided by spatial data and of the techniques under development to take advantage of these opportunities.

Notes

1. Definitions of geographical information systems (GIS) are legion. DeMers (1997, p. 7) suggests that they are 'tools that allow for the processing of spatial data into information, generally information tied explicitly to, and used to make decisions about some portion of the earth'. We examine the role of GIS for spatial data analysis in Chapter 3.
2. Flat sheets of paper can be bound into reports and atlases, and the flat display of a computer screen provides an analogue to the sheet of paper for ephemeral displays of geographical data.
3. The Public Use Microdata Samples (PUMS) are a welcome step forward in permitting the nature of the aggregation problem to be examined. In the UK, the Office of National Statistics has produced an equivalent, the Sample of Anonymised Records (SAR).

3 The Role of Geographical Information Systems

3.1 Introduction

This chapter is concerned with the role of GIS in the analysis of spatial data. It is not, however, a primer on GIS; rather it deals with the way we use GIS for the advancement of quantitative spatial analysis. Some knowledge of GIS is assumed and the reader is referred to current texts such as DeMers (1997), Burrough and McDonnell (1998), Martin (1996) or Heywood et al. (1998), for further details of the principles underlying GIS. A useful guide to some of the computing issues germane to GIS is provided by Worboys (1995).

The following section illustrates some simple types of spatial analysis using data for an area in North-East England; such manipulations are common in GIS. The third section considers the use of GIS as a data integration and manipulation tool in a modelling framework. The fourth section examines some problems that can arise from the naive use of GIS in spatial analysis. The fifth section considers some alternative approaches to linking other analytical software with GIS, and the chapter finishes with a brief consideration of the future role of GIS in the analysis of spatial data.

In this chapter we shall make reference to the Arc/Info GIS.[1] This is not an endorsement of Arc/Info as a GIS package, neither is it a critique of Arc/Info; it merely reflects the authors' familiarity with this package, and its availability in many universities in the UK and the USA. However, many of the GIS operations to be described in this chapter are available in one form or another in other GIS systems; thus the comments that follow are intended to be general, rather than particular to one software system.

The term 'geographical information systems' can be confusing. At one extreme, it is often used to describe a piece of software, such as Arc/Info, or SPANS. Such software is intended to be general purpose by providing a means of storing, querying, integrating, retrieving, displaying and modelling spatial data. They are characterized by a command language or menu system of some complexity and considerable skill is usually required to operate them. In one sense, such packages are the GIS analogue of statistical packages, such as SAS or SPSS, which are collections of statistical routines that can be applied to a wide variety of data. With GIS, as with statistical packages, it is up to the user to decide which are the appropriate operations for data management and analysis in any given situation. Merely because some technique is provided in a package, does not imply that it is of universal applicability. At the other extreme, GIS might be a specifically tailored, turnkey system for aiding such activities as traffic management, automated mapping or facilities management, as is often found within the offices of public agencies. A predetermined subset of operations (perhaps prepackaged behind a graphical user interface) will be available for operators. There will be a well-defined means of data entry and the operations will be highly specific to the task for which the software has been designed.

What is the role of GIS in the quantitative analysis of spatial data? The latest thinking and ideas on GIS are contained within the academic literature in journals such as the *International Journal of Geographic Information Science* and *Transactions in GIS*. Inevitably, there will be a lag between the description of a new method, or data structure, in the scientific literature and its appearance (if at all) in a particular software vendor's product. Software vendors are not in the business of selling experimental software to customers; you are more likely to find some tried and tested techniques in commercial software, rather than the latest methods for, say, pattern analysis. Indeed, it is not unknown for academics, frustrated with what is available commercially, to make experimental software available. A notable example here is SPLANCS (Rowlingson and Diggle, 1993), a collection of S-Plus routines for the analysis of spatial data produced by researchers at Lancaster University, England.

What can be achieved with the operations programmed into the system by the manufacturer, and what can be done when some other technique is required? Vendors often claim their system has 'advanced spatial analysis functionality'; what do they mean by 'spatial analysis', and does this accord with our own view of spatial analysis? Johnston (1994) defines spatial analysis as the 'quantitative procedures employed in locational analysis'; for Johnston, the focus of locational analysis is 'the spatial arrangement of phenomena' (1994). Indeed Unwin's text *Introductory Spatial Analysis* (1981) is largely concerned with the arrangement of points, lines and areas in space. Johnston suggests that practitioners of spatial analysis have either concentrated on applying linear models to their data, or sought new methods of dealing with the problems brought about by attempting to analyse spatial data (see e.g. Haining, 1990; Bailey and Gatrell, 1995). For DeMers (1997), spatial analysis encompasses a wide variety of operations and concepts including simple measurement, classification, surfaces, arrangement, geometrical overlay and cartographic modelling. Martin's (1996, p. 59) view of 'fundamental operations of GIS' includes reclassification, overlay, distance and connectivity measurement, and neighbourhood characterization. Clarke (1997, p. 184) identifies a number of 'geographic properties' of phenomena, including size, distribution, pattern, contiguity, neighbourhood, shape, scale, or orientation, and asks the question: how can variations in these properties be described and analysed? If some of these points of view represent ideals, what are we likely to find? When GIS software first became widely available, many academics were surprised to find that the term 'spatial analysis' had been borrowed by GIS vendors to describe what were essentially a series of geometrical operations. Little of the stochastic framework that had characterized much of the spatial analysis approach was present in GIS. What do we then find, loosely characterized as spatial analysis, in a typical software product?

3.2 Simple GIS-based spatial analysis

One characteristic perhaps differentiates GIS from other software that is intended merely to map layers from a spatial database. This is the ability of the former to calculate and store spatial relationships between different features in the database, either within the same layer, or between layers. An early example was in the development of an algorithm which would determine whether a point lay inside a polygon or not. The underlying problem is one of computational geometry and a number of solutions have been proposed. One commonly used solution is to construct a line from the point in one direction. If this line crosses the digitized boundary of a polygon an odd number of times, then it lies inside the polygon, and if it crosses an even number of times, the point lies outside the polygon. One has to make some decisions in special cases, for example when the point in question lies on the boundary. Sedgwick (1990, p. 353) provides code in C for this algorithm, and Baxter (1976, p. 170) provides a FORTRAN IV example.

In order to illustrate some of the GIS functionality claimed as 'spatial analysis', we shall use some digital data for a small part of North-East England. The data are extracted from a much larger database created by Bartholomew (Bartholomew, nd) sourced from 1:625 000 material. As well as the Bartholomew data, we shall take some data from the 1981 UK Census of Population at ward level.

3.2.1 Feature selection by attribute

One of the layers in the Bartholomew data is the boundaries of the administrative counties of England and Wales, and the administrative regions of Scotland. These boundaries are statutorily defined, and refer to the administrative geography of the UK on 1 April 1974. The features in this layer are polygons, and one of the attributes of each polygon contains the name of the county which that polygon represents. To demonstrate a simple GIS-based spatial operation, we can create a new coverage based on one or more attributes of the original coverage. Here, for example, we create a new coverage by selecting one polygon from the UK database, that of County Durham (see Figure 3.1). Many GIS have a function which permits the selection by features on the basis of attribute value. The input here is the UK county polygon coverage, and the output is a polygon coverage containing a single polygon representing the boundary of the administrative county of Durham. The county is about 70 km from east to west, and about 50 km from north to south. Its area is about 2400 square kilometres.

Figure 3.1   **Feature selection by attribute**

3.2.2   *Feature selection by geometric intersection*

Among the layers in the Bartholomew data are those representing the positions of
the railway lines, and those representing the extents of the urban areas. The railway
coverage is a line coverage, and the urban area coverage is a polygon coverage. To

extract those features which lie within the boundary of Durham county requires a different approach to that in the last section. Neither coverage has an attribute that contains the name of the county in which each feature lies. Most GIS will supply a function to allow clipping or 'cookie-cutting'. One coverage is designated as input, and the boundaries of another are taken as a clipping coverage. The output is a coverage whose features lie within the extent of clipping coverage. Figure 3.2(a) shows the county boundary with the urban areas from the complete Great Britain coverage, and Figure 3.2(b) shows both the urban areas and railway lines after the clipping operation.

3.2.3 Buffering features

A commonly encountered facility in GIS is the ability to create a buffer. This involves the creation of a circular region around a point, or a corridor around a line feature. The radius of the circle, or the width of the corridor, is defined by the analyst. Some systems permit flexibility in specifying variable buffer widths (to reflect varying properties of the features). Figure 3.3 shows a 1 km buffer for the railway lines in Durham. This might be used to reflect notions of accessibility, or perhaps local susceptibility to noise from rail transport operations. The output from a buffer is always a polygon coverage.

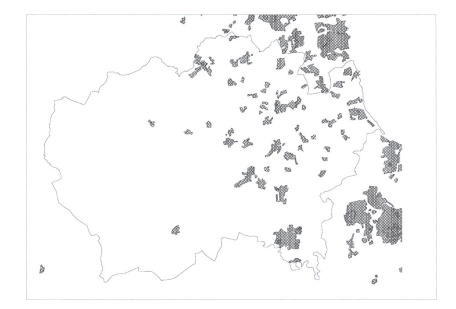

Figure 3.2(a)   **County boundary with unclipped urban area boundaries**

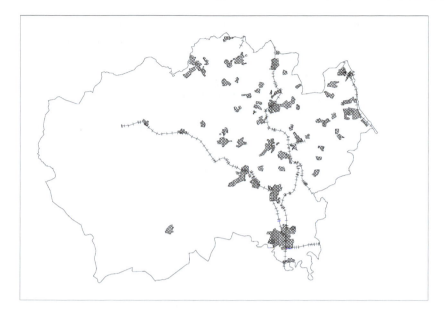

Figure 3.2(b)   **County boundary with clipped urban areas and railway lines**

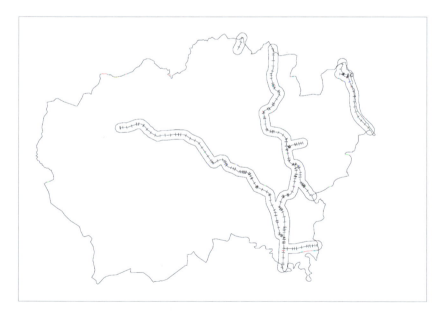

Figure 3.3   **A 1 km buffer around railway lines**

3.2.4   Geometric intersection: union

If we wish to calculate the proportion of urban area within the rail buffer, then we shall need to create another coverage containing both the rail buffer and the urban areas' extents. The appropriate operation here is a union of the two spatial data sets. Where a polygon is crossed by the boundary of the buffer, two new polygons will be formed, one outside the buffer and one inside the buffer. To display the parts of the buffer in the area, we will need to inform the software that only those polygons which were *both* (a) originally part of the urban cover *and* (b) within the buffer polygon are to be selected for shading. Figure 3.4(a) shows the combined

Figure 3.4(a)   **The union of the 1 km rail buffer and the urban area**

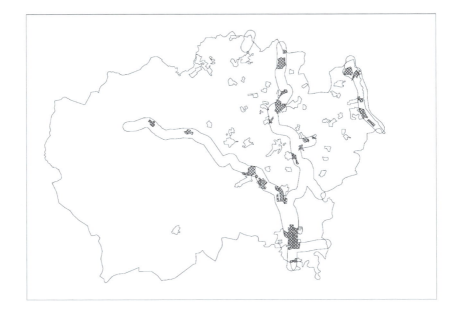

Figure 3.4(b)   **Urban areas within the rail buffer**

polygons created from the union of the urban area boundaries and the rail buffer. The shaded polygons in Figure 3.4(b) are those which satisfy the criterion above. Additionally we can then calculate the area of urban land within and without the buffer: in this example, of the 360 square kilometres of urban land, about 40 square kilometres lie within 1 km of a railway line.

3.2.5 Geometric intersection: intersection

If we wish to create a new coverage of only those areas which are both urban and in the rail buffer, an intersection operation will be required. This is a slightly different operation to the selection of areas for display from the union operation in the previous section (which was based on selection by attribute value). Intersection is an operation that may be carried out on polygon, line or point data, the output coverage being of the same type as the input coverage. Figure 3.4(b) shows the urban area polygons that lie in the rail buffer as shaded areas.

In Figure 3.5(a) the railway lines that lie in the intersection of the railways with the urban areas have been omitted from the display. Figure 3.5(b) shows the output of an intersection of the ward centroids with the urban area buffer (allowing us to calculate the number of residents living within the urban areas). Ward centroids lying within an urban area extent are shown as black squares, and those outside are shown with lighter symbols.

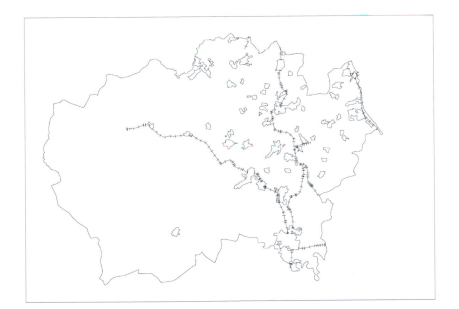

Figure 3.5(a) **Railway lines outside urban areas**

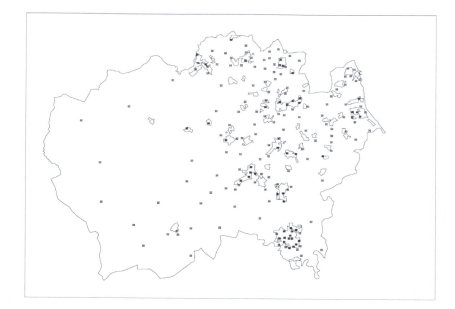

Figure 3.5(b)   **Ward centroids in urban areas**

3.2.6   *Proximity*

A related operation is the computation of the proximity of objects in one layer to objects in another. Typically, one layer will be point data, and the other layer will be point or line or polygon data. The output from such an operation is the addition of at least two new attributes to the attribute table for the input layer (attributes in the vector model are outlined in Chapter 2). The first will be the ID of the nearest object in the other layer (nearest point, or nearest line segment perpendicularly) and the second will be the distance in whatever units are being used. This operation can be time consuming if inefficiently programmed; if there are n objects in the first layer, and m in the second, nm distance calculations may be required. This operation may appear to be of limited utility. However, with a point layer representing the locations of settlements and a line layer representing roads, populations can be assigned to road segments (perhaps inversely weighted by distance) for network modelling applications.

A related operation in several GIS packages is the creation of Thiessen polygons from irregularly spaced point data. The boundaries of Thiessen polygons are created such that any location inside a polygon is closer to that polygon's centroid than to any other polygon centroid. Thiessen polygons have a characteristic convex shape. As with buffering, the output from this operation is a polygon coverage. However, there will be as many polygons in the output layer as there are points in the input layer (this may not be the case with a buffer operation on point data). For points on the edge of the study area, the boundaries of Thiessen polygons tend to form exaggerated areas (usually they are far too large). A possible solution is to clip the layer with a digitized boundary to remove the worst of the problem polygons. Thiessen polygons can be useful for producing choropleth maps from point data. In reality many boundaries tend to exhibit both local convexities and concavities; as an example, a boundary which follows the path of a meandering river will be alternately convex and concave. The boundary between the administrative districts of Newcastle and Gateshead, which follows the path of the River Tyne, is such a boundary, as is the boundary between the states of Ohio and West Virginia, which follows the Ohio River.

Figure 3.6(a) shows the Thiessen polygons created by the Arc/Info software for the ward centroids. Notice that the boundaries of the polygons extend beyond the county boundary. In Figure 3.6(b) the Thiessen polygons have been clipped with the county boundary. In the absence of a set of digitized boundaries, Thiessen polygons are often used as pseudo-boundaries; Figure 3.7 shows a map of population density created from the Thiessen polygons. The data are classed into six groups, such that there are approximately the same number of wards in each group. The map is shown with the boundaries drawn (Figure 3.7(a)) and without (Figure 3.7(b)). A larger number of grey shades have been used in the lather to give an indication of shading without the imposition of classes. The contrast in density between the rural areas in the west of the county and the more urbanized areas in the east is noticeable in these two maps.

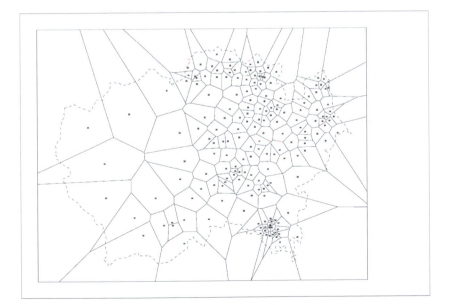

Figure 3.6(a)   **Thiessen polygon boundaries for Durham wards**

Figure 3.6(b)　**Thiessen polygons clipped into the county boundary**

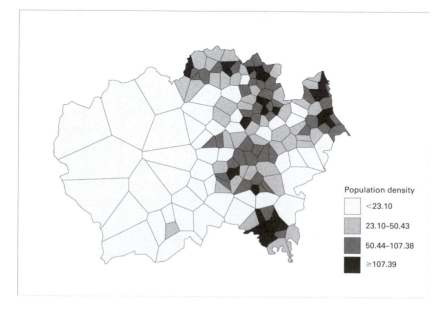

Figure 3.7(a)　**Classed population density by ward**

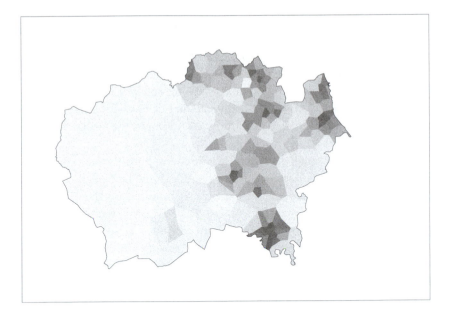

Figure 3.7(b)   **Population density by ward – continuous shading**

3.2.7   Contiguity

A by-product of the creation of a topologically correct polygon layer can be a table showing, for each arc in the layer, the IDs of the polygons on either side of the arc. Such a table can be useful in the dissolve operation. Dissolving involves removing boundaries from a layer which border two polygons possessing the same values for one or more attributes. The output is a new polygon layer with the same number or fewer polygons than the input layer. This operation is useful with administrative data in which there is a spatial hierarchy. If one has data for the lowest level in the hierarchy, layers for the other levels can be built, if the hierarchy is represented in one or more of the attributes in the data.

In the spatial hierarchy used by the UK Office of National Statistics, the lowest levels at which the decennial population census data are reported are called enumeration districts. These nest into larger units known as wards, which in turn nest into administrative districts, which nest in turn into administrative counties. Each enumeration district is given an eight-character code; the first two characters identify the county; the next two identify a district; the third to sixth inclusive identify a ward; and the third to eighth identify an enumeration district. Given a layer of digitized enumeration district boundaries, it is possible, using a series of dissolve operations, to recover the digital representations of the other areas in the hierarchy. As an example, county 06 is the code for the county of Tyne and Wear; 06CJ is the coding for the administrative district of Newcastle upon Tyne; 06CJFA refers to the ward known as Jesmond (a suburb of Newcastle); and 06FDFA01 is one of the enumeration districts in this ward. Jesmond is one of 26 wards in

Newcastle upon Tyne, and there are five administrative districts within Tyne and Wear.

In the USA, the FIPS (Federal Information Processing Standard) assigns codes for both states and counties. For example, the FIPS code for New York State is 36, and the Code for Erie County is 029 (one of the 62 counties in New York State). The five-digit code 36029 combines both state and county levels and provides a unique identifier for Erie County. As with the UK example, the spatial hierarchy is mirrored in the county code.

Both the above are examples of fixed length codes. For the county FIPS code, the number is padded on the left with zeroes to give 3 digits. Not all codes which are hierarchical are fixed length. An example is provided by the UK postcode system. The postcode for Newcastle University is NE1 7RU; the part of the code up to the space is known as the 'outward' part, and the last three characters form in 'inward' part. The 'inward' part is always digit–character–character. However, the outward part specifies, first, the post town (characters), and then a postal district within this area (digits). Not all post towns are identified by two letters (NE is 'Newcastle' but Sheffield is 'S' and Glasgow is 'G'). There may be more than nine postal districts within a postal area, but those with codes 0 ... 9 are not padded with a zero to give two characters. This may cause problems when parts of a code are being extracted for a dissolve operation.

The dissolve operation can be used with some success following the creation of Thiessen polygons for some small spatial units to generate boundaries for areal units that are part of the same hierarchy. Although the atomic units will be convex, those higher-level units may exhibit reasonably plausible concavities, although they are no substitute for the actual boundaries, and may even contain contiguities which would not exist in reality.

3.2.8 Interpolation and fields

A mathematical model for representing the variation of some phenomenon over a continuous expanse of geographical space is sometimes referred to as a field. An example of a field is the spatial distribution of soil pH values. As computers cannot handle continuous functions directly, some form of discrete approximation must be used. Given a spatial framework, for example a partition of the area of interest into tessellating squares, a field can be modelled as a computable function from the spatial framework to an attribute domain. Usually the spatial field gives the z values at locations on a horizonatal $x-y$ plane (Worboys, 1995, pp. 152–3). One means of visualizing the field is as a surface, either as a set of isolines (e.g. contours), or in pseudo-3D. Field-based models are usually implemented in raster GIS. In Arc/Info the GRID module is available for handling field-based data, although there are standalone packages, notably IDRISI. We shall use the terms grid and surface interchangeably below to refer to layers in a field-based model.

Some systems permit the creation of a surface (see Chapter 2) from irregularly spaced point data (or line data). This requires a set of points each of which has an associated z value from which the surface will be interpolated. There are a number of alternative techniques for creating grids, including inverse distance weighting, splines (see Chapter 4) and kriging (see Chapter 7). However, the user needs to take some care in surface interpolation from such data, since most techniques require that some parameters are set to control surface generation. Injudicious choice of parameter values may give rise to a surface that is very different from that which was intended. A common use of such interpolation techniques is the creation of a digital elevation model from digitized contours and spot heights. However, the z value that one uses in interpolation need not be elevation, it might be some other attribute which varies spatially, such as population, or unemployment rate, or the parameters from the calibration of a localized model (see Chapter 5). Figure 3.8(a) shows the variation in elevation in the landscape of Durham – upland areas are shaded a lighter grey than lowland areas. It is instructive to compare this map with the population density maps in Figure 3.7. The more urbanized, densely populated, areas are in the lowlands on the east side of the county. Figure 3.8(b) shows an attempt to visualize the surface in pseudo-3D. A mesh, or wire frame, is 'draped' across a representation of the surface, such that valleys and hills can be clearly seen. In order to enhance the impression of elevation change, the heights have been multiplied by 9. The choice of vertical exaggeration is largely a matter of experiment.

An alternative perspective on the creation of surfaces is provided by Tobler's smooth pycnophylactic interpolation (Tobler, 1979). Here the goals are to create a smooth surface and to preserve the total value of the attribute being interpolated

Figure 3.8(a) **Shaded digital elevation model**

Figure 3.8(b) **Surface represented as a wire frame**

within each source zone. This method starts with a rasterized version of a zonal partition of the study area, with zones z_1 to z_m. If the ith zone contains n rasters, then each raster in that zone is initially allocated $1/n$th of the value of the attribute being interpolated. The surface is then smoothed, using a low-pass filter.[2] The resulting smoothed values for the ith zone are summed, and their values adjusted uniformly such that their sum is the same as the original attribute total for that zone. All zones have their values adjusted accordingly. The smoothing, summation and adjustment process is repeated until there is no change between successive iterations.

The ability to represent data as a surface has some advantages from the modelling point of view. Tomlin (1990) coined the term 'cartographic modelling' for analysis using grids. Some of Tomlin's ideas are included in the modelling language used in the GRID module of Arc/Info. Analysis takes place by creating grids, or by combining or modifying other grids. Grids may be added, subtracted, multiplied and divided; they may be compared and logical operations (such as AND and OR mentioned above) may be carried out. All these operations are done on a cell-by-cell basis. Grids can also be modified by using functions; in the GRID module, functions may be 'local' (working on a cell-by-cell basis), 'focal' (including neighbouring cells), 'zonal' (neighbourhoods are defined by another 'zone grid'), or global (relating to the whole grid). As examples, the local function sqrt returns an output grid in which the cell contains the square root of the corresponding cell in the input grid. The focal function focalmean returns an output grid in which each cell contains the mean of a spatial filter whose shape is under some user control. The global function costdistance returns an output grid in which each cell contains the cumulative cost–distance to some location; the user also supplies a 'cost' grid which contains the unit 'cost' of travel through each cell. Other commands permit the conversion between vector representations and the grid representation; conversion from grid back to vector is also possible.

3.2.9 Density functions

Some GIS allow the estimation of a density surface from irregularly spaced point data. Given an input layer representing the locations of a discrete set of events, one can create an output grid in which each cell contains an estimate of the spatial density of those events. Among the usual methods available for density estimation is one employing a spatial kernel (see also Chapters 4, 5 and 6 and Silverman, 1986).

One variety of density estimate for a distribution is provided by a histogram. Given an origin and bin width (in the units of the data we are graphing), we can construct a histogram quite easily. The height of the bar for any one bin gives an indication of the density of observations in that bin relative to the other bins in the histogram. With spatial data, we can create a simple density estimate if we extend the histogram into two dimensions; in other words, its base becomes a grid with known origin and bin (or cell) size. Such a histogram can be visualized in many software packages, for example in pseudo-3D, or by using a greyscale to indicate variations in density. In both cases, larger bins will in general produce 'smoother' histograms than smaller bins. However, it is not always clear what bin size to choose. Also, the histogram imposes an arbitrary quantization on the original data if they are continuous.

The technique of density estimation allows us to place a *kernel* (or hump) over each observation. The kernel can be thought of as a generalized bin. Summing the individual kernels gives us the density estimate for the distribution. An estimate of the relative density of a distribution in two-dimensional space around a location (x, y) can be obtained from the following:

$$\hat{f}(x, y) = \frac{1}{nh^2} \sum_{i=1}^{n} K(d_i / h) \qquad (3.1)$$

where $f(x, y)$ is the density estimate at the location (x, y); n is the number of observations; h is the window width (sometimes referred to as the smoothing parameter or bandwidth); K is a kernel function; and d_i is the distance between the location (x, y) and the location of the ith observation.

The kernel function K defines the shape of the humps to be placed over the individual observations, and the bandwidth controls their widths. The resulting density estimate is smooth, and is a probability density. The aggregate effect of placing humps over each observed point is to create a continuous distribution reflecting the smoothed variations in the density of the point data set. Note that we may obtain an estimate of the density anywhere in the study area and not necessarily at the locations where the observed data have been sampled. Typically we may wish to compute $f(x, y)$ at the mesh points of a rectangular grid.

The analyst has several choices to make. The first is to decide on the appropriate kernel shape. Silverman (1986, p. 43) provides formulae for some common kernels but points out that there is little to choose between different kernels on the basis of their ability to replicate the shape of the distribution. The shapes of some common kernels are shown in Figure 3.9.

The selection of the bandwidth h controls the amount of smoothing of the data: larger bandwidths will tend to highlight regional patterns, and smaller bandwidths will emphasize local patterns. Density estimates generated in this fashion are preferable to those generated by aggregating data to a grid as the arbitrary quantization imposed by the grid is removed from the observed pattern.

Figure 3.10 shows the output from the pointdensity function in GRID using the residential population at each ward centroid as input data. This is somewhat similar to the population surfaces of Bracken and Martin (1989). In Figure 3.10(a) a bandwidth of 2 km is used, whereas in Figure 3.10(b) a bandwidth of 5 km is used. Notice the difference in the variation in apparent population density that occurs. With the larger bandwidth, the greater is the smoothing of the data. It is instructive to compare this map with that in Figure 3.7(a), in which an artificial set of boundaries has been imposed on the spatial distribution. The latter erroneously encourages the impression that population density is not a continuously varying phenomenon.

The incorporation of these more advanced data manipulation techniques in GIS is welcome, as it permits some interesting analyses (Kelsall and Diggle, 1995). One might, for example, have two sets of irregularly spaced data, the first representing the locations of individuals with some disease, and the second representing the locations of an at-risk population for the disease. If the disease is not spatially clustered, then we would expect the incidence of the disease to be uniform in the population. Two density grids may be created, one for the individuals with the disease, and one for the at-risk population. If the diseased population density grid is divided by the at-risk population grid, we obtain a new grid showing the disease incidence (per capita incidence). This can be mapped to examine variations from the overall disease incidence (the total number of disease cases divided by the total at-risk population). Further tests can be undertaken to determine whether or not local variation is outwith some confidence limits for the data.

3.2.10 Analysis on networks

A frequently encountered group of functions provides the ability to model paths through networks. Typically such networks are road networks and the analyst is concerned with finding a path from one location to another which is 'optimal' in

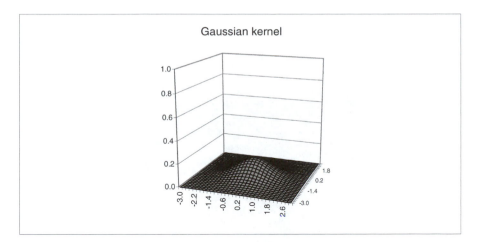

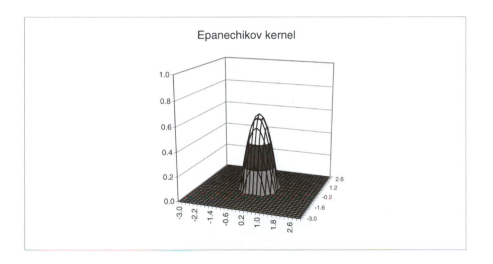

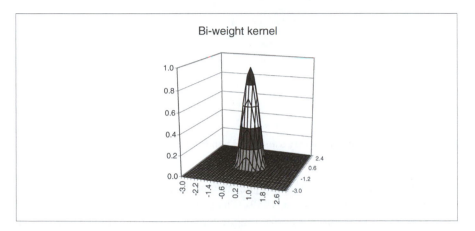

Figure 3.9   **Kernel shapes in three dimensions**

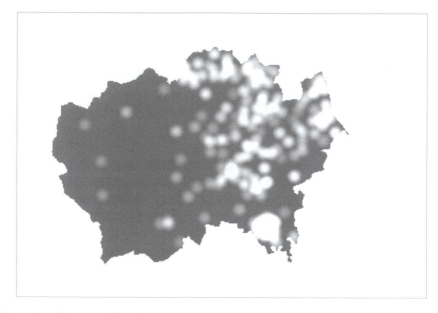

Figure 3.10(a)   **Population density using 2 km bandwidth**

Figure 3.10(b)   **Population density using 5 km bandwidth**

some sense. The algorithms for carrying this out are usually based on Dijkstra's shortest-path algorithm (Dijkstra, 1959). The user will have to supply the attribute to be used for finding the shortest path; if arc length is used, the resulting route will be the shortest. If times can be supplied for passage along each arc in the network, the resulting path will be the quickest; if passage costs are available, the resulting path will be the cheapest. Variations on this include being able to model the shortest route between a variety of locations, either in a specified order, or in an order to be determined by the network analysis function. Some systems permit other functions, such as allocation, in which arcs are identified as being reachable in some specified distance or time from a given location.

The creation of optimal routes has spurred a spate of software from suppliers not traditionally associated with GIS. Perhaps the best known of these is Microsoft's AutoRoute Express product; this provides a user-friendly interface to the shortest-route problem. Once the user has specified a series of places to be visited, AutoRoute Express calculates the shortest route and displays the results on a map, as well as providing directions in printed form for a driver.

3.2.11 Query

Perhaps one of the most basic of all GIS-based spatial analytical tasks is the ability to query a coverage interactively. This is usually achieved by pointing to a location on the on-screen map display, and pressing a button on the mouse to request the retrieval and display of the attributes associated with the feature which has been selected. In some systems, the user may use one map to guide where the query takes place, and be able to query another layer or layers simultaneously. As an example of the latter, we might be interested in the variation of population density relative to variation in topography. Figure 3.11 illustrates querying the population density surface in the vicinity of the centre of Darlington – the user sees and is guided by the map of urban area extents and the location of the railway line, but queries an unseen surface.

Query is also possible by attribute, allowing the user to perform such tasks as selecting and displaying all zones with a population density greater than 2.5 persons per hectare. This may be followed by a spatial query, with the analyst querying either the values of the attributes in the zones highlighted, or another layer in the database.

3.3 Advanced GIS-based spatial analysis

GIS can be considered to have three basic roles: as a data manipulator; as a data integrator; and as a data explorer. We have considered the role of GIS as a data manipulator in the above section that covers the basic spatial data operations available in most GIS. Here we consider the role of GIS in more advanced forms of GIS-based spatial analysis that tend to involve data integration and data exploration.

Fotheringham (1999b) considers the position of GIS and spatial analysis from two viewpoints. The incorporation of simplified versions of established modelling frameworks, such as location–allocation models or spatial interaction models, can

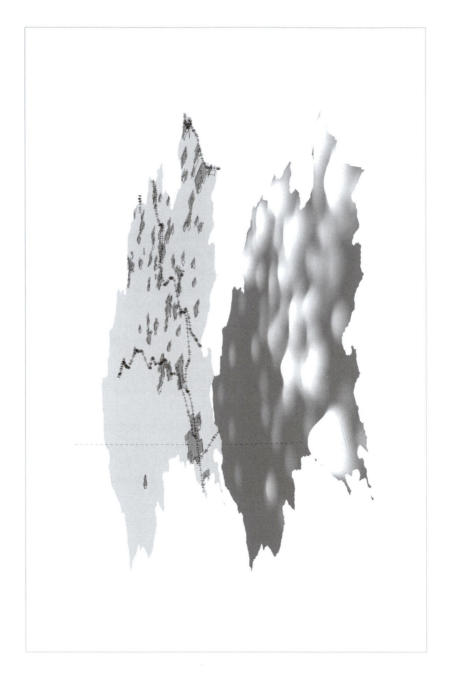

Figure 3.11 Querying the population density surface from the topographic map

be regarded as a retrograde step. The experience which GIS have engendered is that spatial problems are far from easy, and the inexperienced user is likely to be misled by the apparent simplicity with which a solution to, say, a *p-median* problem can be produced. On the other hand, the use of GIS with spatial models, where GIS have acted as a catalyst to a new modelling form, can be regarded as progress. In the mid-1980s the widespread diffusion of GIS caused a resurgence of interest in spatial problems amongst geographers, and, perhaps more importantly, researchers in other disciplines. Part of this catalysis was provided by the functionality of the software itself but an important part has been provided by institutional support for GIS, particularly in universities. The establishment of the National Center for Geographic Information and Analysis in the USA, and the Regional Research Laboratories initiative in the UK, were notable in this regard.

Fotheringham (1999b) identifies three examples where the linkage of spatial modelling and GIS can be regarded as a step forward. One area has been interest in modelling local variation, either in attributes or as a result of modelling (see Chapter 5). A second area is concerned with examining the sensitivity of results of a location–allocation modelling exercise with respect to changes in the aggregation of the data used in the model (see Chapter 10 and Fotheringham et al., 1995). The third area is concerned with computer-intensive point pattern analysis, an approach pioneered by Openshaw et al. (1987). In each case, GIS play a role as a data integrator in the pre-modelling phases and as an evaluation tool in the post-modelling phase. In no case can GIS be regarded as central, but they are extremely useful as a facilitator. Indeed, in Openshaw et al.'s work, the 'GIS' amounted to hand-coded routines for the fast retrieval of spatially referenced data and some separate routines to send the results to a pen plotter.

3.3.1 Data-integration and management

Perhaps one of the most useful benefits that GIS packages offer is that of data integration and management. Data for a modelling application may come from a variety of sources. Powe et al. (1997) report the use of GIS in assisting with the assemblage of data for an exercise designed to assess the effects of proximity to woodland on house prices in the New Forest area of southern England. Their modelling methodology was based on the *hedonic* model, in which variation in individual house prices is related to the attributes of the properties under study. Typically the attributes may include:

- Structural characteristics
- Socio-economic characteristics of the neighbourhood
- Aesthetic and environmental characteristics of the neighbourhood
- Geographical characteristics
- Property rights

Data for such modelling can come from a wide variety of sources. The structural characteristics can be taken from building society mortgage records (some are available in the UK; Clarke and Herrin (1997) used a data set from TRW REDI-Property). Socio-economic information concerning the neighbourhood of the property can be obtained from Census of Population data. Aesthetic and environmental characteristics, for example proximity to watercourses, or woodland, may have to be captured from existing printed maps or aerial photographs and some GIS processing is likely to be required. The geographical characteristics of the property, such as accessibility to urban facilities, may have to be determined through some GIS processing. In some countries, a property cadastre may be a valuable source of information on property rights.

Powe et al. (1997) specified a price function of the following form:

$$P_i = f(\mathbf{AM}_i, \mathbf{ENV}_i, \mathbf{S}_i, \mathbf{SE}_i, Y_i) \tag{3.2}$$

where P_i is the price at which the house was sold; $\mathbf{AM}_i$ is a vector of accessibility and locational variables; $\mathbf{ENV}_i$ is a vector of environmental variables; $\mathbf{S}_i$ is a vector of structural characteristics of the ith property; $\mathbf{SE}_i$ is a vector of socio-economic variables for the surrounding ward; and Y_i is the year in which the ith property was purchased. The particular variables chosen for the model are listed in Table 3.1.

Each property in the building society's database was given a grid reference and one layer of the structural characteristics of each property was constructed using Arc/Info. It should be remembered that building societies do not generally store their data in GIS, so some extra processing was required to extract from the mortgage records only those variables that were of interest to the study. Some sources of data which come the way of the spatial analyst may have some form of geo-reference already attached (e.g. a postcode or zip code). If this is the case, a suitable lookup table may be used.

In order to obtain the woodland variables, the identification of woodland (with a car park or picnic area) from maps and the digitizing of the woodland boundaries was necessary. For each property the distance to the nearest woodland and the distance to the New Forest Park were obtained using the near command in Arc/Info. A dummy variable was also created to indicate whether a property was within 500 m of a woodland boundary. Properties were also intersected with the New Forest Park boundaries and those inside the New Forest Park allocated a dummy value to indicate this location.

It is not surprising that Powe et al. (1997) found significant correlations between the various woodland variables created using GIS. In order to have a single variable that would act as a proxy for woodland characteristics, they created the following index for each property:

Table 3.1   **Variables used in Powe et al.'s (1997) hedonic modelling study**

Type of variable	Description
1 Woodland variables	
Digitized data	Distance to nearest woodland
	Distance to New Forest Park
	Dummy: location within 500 m of woodland
	Dummy: location in New Forest Park
	Woodland index
2 Other amenity characteristics	
Digitized data	Distance to sea
	Dummy: location within 500 m of sea
	Dummy: location within 200 m of a river
	Distance to nearest large urban area
	Dummy: location in large urban area
	Dummy: location within 500 m of an oil refinery
	Dummy: location within 100 m of a railway line
	Dummy: location within 100 m of a main road or motorway
3 Structural characteristics	
Building society data	Floor area
	Number of bathrooms
	Number of bedrooms
	Dummy: detached building
	Dummy: semi-detached building
	Dummy: terraced building
	Dummy: garage
	Dummy: full central heating
	Age of property
4 Socio-economic data	
1991 Census of Population	Proportion of residents aged below 18
	Proportion of families with no car
	Cars per person
	Proportion of employees in professional occupations
	Proportion of employees in unskilled occupations
	Proportion of retired employees
	Male unemployment rate
	Unemployment rate

$$\text{Forest access index}_i = \Sigma_j(\text{area}_j/\text{distance}_{ij}^2) \qquad (3.3)$$

in which area represents the area of the jth woodland and distance is defined as the distance to the nearest point on the boundary of that woodland.

The creation of the other amenity variables again required digitization of the coastline, rivers, large urban area boundaries, oil refineries, main roads and motorways. Further processing within GIS was necessary to create the relevant distance and dummy variables in the same manner as described for the woodland variables above.

Data from the 1991 UK Census of Population is available at a number of spatial scales. While the concept of 'neighbourhood' is not well defined, Powe et al. (1997) extracted data at the enumeration district level. The average population of an enumeration district is about 440 residents, which represents some 175 house-holds. The UK census data at Manchester University (*http://www.midas.ac.uk*) are not stored in GIS. Powe et al. (1997) do not state how they manage to link the census data for the enumeration districts in their study area with the properties. There are a number of options. Digital boundaries are available (Ordnance Survey, nd) for enumeration districts – the property records may be intersected with these and each property record tagged with the socio-economic characteristics of the enumeration district in which it lies. If no digitized boundaries were available, then another option would be to use the 'centroid' which appears in the census records to calculate, as with the woodland and amenity layers, the closest centroid to each property; then the socio-economic data would be aligned to the property records. A third option would be to use the ONS Postcode-to-Enumeration District lookup table (Dale and Marsh, 1993) to link the property and census data.

Why did Powe et al. use GIS? They comment:

> GIS provides clear advantages. ... Once the spatial data is in a usable format, it is a simple task for the GIS package to derive a spatial dataset for each house. ... Furthermore, the speed and accuracy with which the variables can be generated using GIS, permits a greater variety of spatial variables to be generated.
>
> (Powe et al., 1997)

Later on, they comment:

> The major benefit of using a GIS in this context was that it permitted the location of existing or proposed areas of woodland, in relation to centres of population, to be fully taken into account.
>
> (Powe et al., 1997)

It should be noted that, although Powe and his colleagues used GIS for the integration of a variety of disparate data sources, they did not carry out their model estimation in GIS. The data had to be written to a file, and read into some other statistical software in order to carry out the model calibration. This is typical of the current application of GIS to spatial analysis and spatial models – we have not yet reached the stage where all the required operations can be found in one piece of software.

3.3.2 Exploration

GIS do provide, as well as data integration and data manipulation facilities, a means for exploring the spatial aspects of data. However, few GIS packages provide the type of highly interactive exploration environment which is provided by statistical computing systems such as S-Plus (Venables and Ripley, 1997), R (Ihaka and Gentleman, 1996) or XlispStat (Tierney, 1990). Most GIS packages will allow the user to calculate some simple summary statistics for attributes; these may be used as the basis for exploring attributes singly in the search for values which might be thought 'unusual'. We need to be careful that the confidence limits for positively autocorrelated data will be narrower than if we assumed independence, so any search should proceed with that rider. If the search for outliers reveals the same location or locations repeatedly, this will inform any interpretation we can place on the analysis (see also Chapter 4). Mapping data will also reveal any obvious spatial trends. The user needs to take care in mapping, however, since some of the automatic classing schemes offered by GIS packages may hide spatial trends, so a number of classification schemes should be tried.

3.3.3 Post-modelling visualization

As well as having a role in data exploration in a pre-modelling phase, GIS have a useful role in allowing the exploration of the results *after* modelling. The output from a single run of geographically weighted regression (GWR) (see Chapter 5), for example, may include for each input observation a set of parameter estimates and a set of standard errors. One means of examining the spatial variation in the parameter estimates is to map them. For point data, one mapping technique is to interpolate a surface and then display the surface, either shaded as a greyscale in planimetric form, or in pseudo-3D from a variety of viewing positions. Experience with GWR surfaces suggests that a linear contrast stretch provides a useful means of displaying surfaces as a greyscale. With pseudo-3D mapping GIS provide a variety of possibilities. The standard error surface can be draped over the surface representing the variation in the parameter's values. The parameter surface may be draped over a surface representing the variation in the input data for which the parameter has been estimated. Examples of these may be found in Fotheringham et al. (1996) and Charlton et al. (1996). With area data, an obvious output for a set of geo-coded results is a choropleth map. The choice of class interval for a choropleth map is a problem; if the software allows something akin to a continuous scale over the range of values, this is preferable to using five or six fixed class intervals. Injudicious choice of class interval can mislead as well as illuminate.

Figures 3.12 and 3.13 show one attempt to evaluate the results of a modelling exercise. The results are from GWR (Brunsdon et al., 1998b; Fotheringham et al., 1998), and the figures show various methods of displaying the spatial variation in the parameter estimates. The modelling exercise attempted to link variation in the incidence of limiting long-term illness to unemployment and housing density. Figure 3.12(a) depicts the spatial variation in the values of the unemployment parameter as a greyscale, in planimetric view. Figure 3.12(b) illustrates the same surface displayed as a pseudo-3D wire mesh model, viewed from 30° altitude and an azimuth of 215° from North. The marked differences in the parameter values are clear from this display, as is the relative lack of variation elsewhere in the county. Figure 3.13(a) shows a greyscale surface representing variation in the original variable draped over the parameter surface; the instability is in areas of very low unemployment. Finally,

Figure 3.12(a)   **Parameter variation as a greyscale**

Figure 3.12(b)   **Parameter variation as a draped mesh**

Figure 3.13(b) shows a greyscale display of variation in the incidence of limiting long-term illness. The analyst can contrast and compare these maps to try to understand the nature of the spatial variation in the parameters as a post-modelling exercise to provide information on possible model mis-specification.

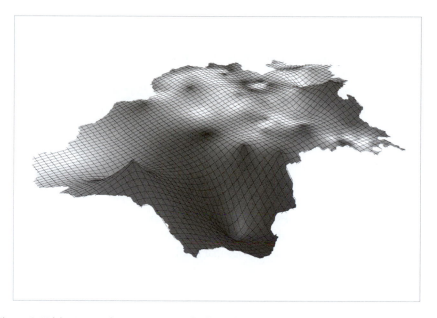

Figure 3.13(a)   **Unemployment greyscale draped over the parameter surface**

Figure 3.13(b)   **Variation in limiting long-term illness shown as a greyscale**

The spatial query possibilities of GIS allow further exploration of the spatial variation in the parameter estimates. One possibility is to display a map of the parameter values and carry out a spatial query on the data layer displaying the values of the dependent and independent variable at various 'interesting' locations, or to query the parameter and data layers simultaneously. It might not be possible to accomplish this with a single command; however, if the GIS software has a macro language, a small macro could be written to accomplish the query after the typing of a single command. If one is accessing the system on which the GIS software is running with a GUI, then the command to run the macro together with its carriage return can be stored on the clipboard for further simplicity of operation.

3.4 Problems

The user needs to be aware that there are several potential problems with GIS-based spatial analysis. The ease with which some operations can be carried out may mislead the user into imagining that the results of such operations are in some sense 'correct'. The following are issues that need to be borne in mind.

3.4.1 Error modelling

Most data used in GIS are sampled. The sampling of coordinates to represent a line depends not only on the decisions made by the cartographer drawing the original line on the map, but also on the decisions made by the digitizer on where to sample the coordinates. The same operator is unlikely to digitize the line in the same way on two separate occasions – anyone who has accidentally digitized a line twice will be aware of this. Similarly, different operators will make different decisions concerning the locations of the coordinates to be sampled. In some polygon overlay operations, representations of the same boundary will appear in different layers. If each layer has been digitized from a different source, sliver polygons are likely to result (typically these are long and thin, and of small area compared with other polygons in the overlay). Burrough and McDonnell (1998, p. 223) identify seven factors affecting the quality of spatial data: currency, completeness, consistency, accessibility, accuracy sources of error in data, and sources of error in derived data and results. In considering accuracy as a factor, Burrough and McDonnell draw attention to the sampling density of the coordinates, positional accuracy of the coordinates, attribute accuracy, topological accuracy, and data lineage. Currently, very few major GIS packages allow the user to view data as being anything other than 'accurate'. However, some researchers are examining whether fuzzy set theory may help in the representation of uncertainty within GIS (e.g. Burrough, 1996).

The point-in-polygon routines in many GIS categorize a point as being either wholly inside or outside the polygon boundary. This ignores any uncertainty that might be present in the digitized representation of the boundary. One possibility is to postulate an error band around the digitized line (Blakemore, 1984). With such an error band there are five states which might describe the relationship between the point and the polygon: (a) wholly inside, (b) probably inside, (c) on the boundary, (d) probably outside and (e) wholly outside.

Much research has been undertaken into the problems of uncertainty in the representation of data in GIS, and the effects of error propagation in multilayered spatial data sets. Openshaw et al. (1991b) describe a procedure based on Monte Carlo simulation to assess the effects of the propagation of data error through a series of polygon overlays. The result is a series of raster representations showing rasters which are present in 95% of the simulations, or 99% of the simulations. Carver and Brunsdon (1994) examine the problem of vector–raster conversion, and its relationship with feature complexity. Using simulated data of known complexity, they found that there is a relationship between line complexity and misclassification error in the rasterization process. Heuvelink et al. (1989) consider the problem of error propagation in modelling with raster data. A useful collection of papers dealing extensively with geographical features which are not conveniently represented as either objects or fields is to be found in Burrough and Frank (1996).

3.4.2 Cross-area aggregation

A temptation with GIS is to use the embedded geometric procedures to estimate data for one set of areal units with data for a different set of areal units. As a general activity this has been referred to as cross-area aggregation. The original data might be observations taken at a series of points (e.g. fire alarm calls, or incidences of particular types of crime or disease), or they might be data provided by a third party aggregated to a particular set of zones. Alternatively, the original data might be zonal data from another source. If our first set of data is zonally based, then a problem arises when the second set of data is recorded for zones with different boundaries. How can we interpolate data from the first set to the second set? A commonly used procedure is to use area weighting, or some third variable if this is available for both sets of zones. Aggregation becomes a simple process of either using weighted averaging, or, if counts are involved, using techniques such as Poisson regression (Flowerdew, 1988).

More recently, Flowerdew and Green (1991) have considered the problem using a technique based around Dempster et al.'s (1977) EM algorithm; one application has been the interpolation of population data between administrative counties and parliamentary constituencies in the UK. The method assumes that there exists a contingency table in which each cell represents an intersection between a source zone and target zone. We know the area of the zone which is represented by any cell, and we know the totals of the rows. However, we do not know the values of the attribute in the individual cells (i.e. they are missing data). The EM algorithm can be used to calculate the values in each cell, and by adding up the columns, the target zone values can be found. There are two steps to the algorithm:

> E: compute the conditional expectations of the missing data, given a model and the observed data

M: fit the model by maximum likelihood to the dataset calculated in the E-step treating the imputed values as 'real' observations

(Flowerdew and Green, 1991, p. 46)

For example, suppose we start with a set of source zones s with values y_s and a set of target zones t with ancillary information x_t. We define the value to be calculated in the intersection of any one source zone and any one target zone to be y_{st}, the area of the intersection being A_{st}. The model for y_{st} is

$$y_{st} \sim \text{Poisson}(\mu_{st}) \tag{3.4}$$

$$\mu_{st} = \mu(\beta, x_t, A_{st}) \tag{3.5}$$

where β represents a set of unknown parameters to be estimated.

In the E-step the conditional expectation of y_{st} is calculated, which is (if the y_{st} are Poisson)

$$y_{st} = \frac{\hat{\mu}_{st} y_s}{\sum_k \hat{\mu}_{sk}} \tag{3.6}$$

The summation in the denominator indicates that the μ are summed over all the zones in the study area.

In the M-step β is estimated in the model $\mu_{st} = \mu(\beta, x_t, A_{st})$ by maximum likelihood. The values of β from the M-step are then used as input to another iteration of the E-step. This process is repeated until convergence. The y_{st} values can be summed over s to give the desired y_t values (Flowerdew and Green, 1991, pp. 46–7).

One advantage of this approach is that it allows a range of extra variables to be included in the interpolation process to improve the accuracy of the result. A GLIM macro, together with code for interchanging data with Arc/Info, has been created (Kehris, 1989). Care is required, however, to ensure that unrealistic assumptions are not being made about the internal spatial homogeneity of the data being aggregated within the source zones. It is not always possible to assume that population is evenly located within a source zone.

Tobler (1991) has pointed out that pycnophylactic interpolation can be used in cross-area aggregation. The process involves interpolating a surface based on the source set of areal units, using the procedure described earlier. The interpolated values for the rasters can then be aggregated to a target set of areal units, as long as it is known which raster belongs to which new areal unit.

3.5 Linking higher-order types of spatial analysis to GIS

It is unlikely that new approaches to spatial analysis will appear in GIS software soon after their publication in the academic literature. For example, Parzen's paper on estimating a probability density function appeared in 1962 (Parzen, 1962) and Silverman's survey of density estimation techniques appeared in 1986 (Silverman, 1986). Yet it was not until the late 1990s that density estimation appeared in one of the mainstream GIS packages. Consequently, in order to link new approaches to spatial analysis and spatial modelling with GIS, it is often necessary to effect such links oneself. There are several ways of doing this:

1   Write the data to a file from GIS, and read it into another program.
2   Transfer the data using operating system calls from GIS into some statistical software.
3   Incorporate new spatial analysis commands into the GIS software.

Option 1 is crude although it is frequently used. The information from the GIS package is written to a file and used as input to the experimental software, and then the results are coerced into some form in which they can be read back into GIS. This can be cumbersome and error prone, particularly if several manipulations need to be carried out so that the data are suitable for the input routines in whatever software is being used. There is also the question of whether the operating environment being used allows painless transfer of data between software packages. For example, if you are working in a Windows environment, and you are using Microsoft Excel 97 as a staging post, you need to be aware that it will hold a maximum of 65 536 rows from a data matrix.

A slightly less crude version of option 1 is to create some add-in macros or menu elements to permit the user to write the data from the GIS software to another program. The spatial analysis can be carried out in the other software. Further macros or menu calls can then be made available to read the data back in, and perhaps display the results from the spatial analysis using the facilities of the GIS package. A good example of this is Anselin's SpaceStat extension for ArcView (Anselin and Bao, 1997). SpaceStat is a standalone program for carrying out various spatial analysis procedures (Anselin, 1988; 1992). The various modules in SpaceStat permit data input and manipulation, creation and manipulation of spatial weights, descriptive spatial statistics, and spatial regression analysis. The ArcView extension permits the transfer of spatial and attribute data to SpaceStat so that the user can carry out analyses not available in ArcView. The results are then transferred back to ArcView and displayed using a customized menu added to those already available in that package.

An example of option 2 is provided by the linkage of ArcView with the interactive exploratory graphics environment Xgobi (Cook et al., 1996; 1997). Xgobi is a software package that runs under Unix and allows the interactive exploratory visualization of multivariate data. Data may be displayed in a variety of ways, including linked scatter- and line plots, parallel coordinate plots (Inselberg, 1988), grand tour and projection pursuit. Moreover, plots may be linked and subjected to interactive brushing (see Chapter 4 for details of these and other visualization techniques). A facility in Unix known as a remote procedure call (RPC) allows separate programs (or processes) to communicate; it is this facility which Symanzik and his colleagues have exploited to allow ArcView and Xgobi to exchange data. As with the SpaceStat extension, an extra menu item appears on the ArcView menu bar to permit various manners of data exchange to take place (these are known as flavours). These include: (a) simple transfer of the attribute data for each spatial location to Xgobi for analysis; (b) a link to display the cumulative distribution function for an attribute; (c) a link to calculate and display variogram cloud plots; (d) a link to display dynamic lagged scatterplots; and (e) a link to create and display multivariate variogram clouds (see Chapter 4). The disadvantage for the PC user is that the software is designed for Unix, and not Windows.

The manufacturers of S-Plus, Mathsoft International, provide a link between S-Plus and ArcView in a Windows environment. This permits the analyst to move spatial data from ArcView to S-Plus for analysis, and then to import the results of any analysis back into ArcView. Additionally, Mathsoft International's S-Plus for ArcView GIS software permits the ArcView user to access a wide range of spatial analysis methods. It is interesting to observe that this is perhaps the first institutionalization of spatial analysis methods with GIS (Mathsoft International have a web page at *http://www.mathsoft.com*).

At the beginning of the 1980s, there was a realization that the available GIS software was woefully deficient in the area of 'spatial analysis'. Some researchers proposed the creation of 'toolkits' of supplementary commands that would provide some of the missing spatial analysis functionality. Two examples are provided by Openshaw et al. (1991a) and Ding and Fotheringham (1992). Both based their toolkits around Arc/Info as a platform. Openshaw et al.'s proposals were for a set of compiled FORTRAN routines to carry out the spatial analyses, together with a set of macros for moving data and computed results between the software and Arc/Info. Ding and Fotheringham's add-on for Arc/Info, the Spatial Analysis Module (SAM), permits the user to access externally written spatial analysis functions, but from within a menu-based interface to the Arc/Info software. Whilst such a module provides extra spatial analysis functionality, the user is constrained to those functions which are supplied, and exploration of alternatives requires programming experience and a suitable high-level-language compiler. As with the other software referred to in this section, it is available on a limited range of machines. The diffusion of add-ons appears to be limited, particularly for systems designed to run on Unix machines, because separate compilations are required for the different manufacturers' systems.

More recently, however, Haining et al. (1998) have made available a set of tools which are designed to work in conjunction with Arc/Info, called SAGE (Spatial Analysis in a GIS Environment). SAGE is more comprehensive than earlier attempts, and provides a variety of tools for data management and creation, data visualization (including linked windows), query, classification and spatial statistical modelling. Amongst the statistical models are relative risk models for disease data, and regression models with spatial error terms. These considerably extend the functionality of Arc/Info for spatial data exploration and modelling.

MacEachren et al. (1998) describe the implementation of exploratory methods in ArcView, with the goal of exploring multivariate health data. The software, written in Avenue (the scripting language provided with ArcView), allows the creation of scalable and linkable scatterplots, linked brushing, highlighting outliers, dynamic classification (class breaks for a distribution) and bivariate mapping. Again, these techniques considerably extend the spatial analysis functionality of ArcView and, being written in ArcView's own scripting language, should function similarly in either a PC-based Windows environment or a Unix implementation.

Option 3 entails adding new spatial analysis and modelling commands to the existing GIS software. This means waiting for manufacturers to include spatial analysis techniques in their software. It is not uncommon, for example, to find an option to determine the shortest path between two nodes on a network, or a solution to the travelling salesperson problem. Different arc attributes may be specified so that the problem can be formulated to find the shortest path, or, perhaps, the quickest path. Options may exist to plot the resulting paths on a map, or to print out the directions for a driver. Simple spatial interaction models are often available (see Chapter 9). However, calibration facilities are crude, and modelling flows along a single row or column in the flow matrix may be difficult if not impossible. A location–allocation modelling option may also be present in the software. The problem with such approaches is that it is sometimes difficult to be flexible, and the user has to rely on the manner in which the software vendors have implemented the techniques in their software. Additionally, the user faces the problem of knowing little of the details of the algorithms that have been chosen. Different algorithms to accomplish the same task may be included in successive versions of the software, perhaps giving different results with the same data.

Another possibility is to ignore GIS altogether, and look to some more specialist software. Bailey and Gatrell (1995) provide a floppy diskette with their text containing the code for INFO-MAP. INFO-MAP is their own software for spatial analysis and modelling. It is not a fully functioning GIS package in the sense of providing a wide variety of spatial data *manipulation* options. It does, however, possess a wide variety of spatial data *analysis* options, controlled through a simple command line interface. Some simple mapping facilities are provided both for the exploration of data and for the display of the results of any analysis where this is appropriate. The analysis options available mirror those described in Bailey and Gatrell's book, and each chapter is provided with some exercises to illustrate the use of the spatial analysis techniques in practice, and to familiarize the user with the software.

3.6 A role for GIS?

Most commercial GIS software is poorly resourced with advanced spatial analysis functionality of the type described in this book. You either have to write your own software, or use someone else's. As has already been observed in this chapter, there is a time lag between an innovation in spatial data analysis taking place and its inclusion in commercial software. To find that one's GIS software does not contain a means of calculating K functions, or that the kernel density estimation functions have no means for choosing a bandwidth, or that the spatial interaction models are oversimplified, is frustrating at best and misleading at worst. For such analyses one is forced to turn to other software. Such software may be something that the analyst codes in a high-level language such as FORTRAN or C. It may be a standalone package of spatial analysis functions, such as SpaceStat. It may be an interactive, statistical/graphic environment such as LispStat, or R.

However, GIS do have a role to play in spatial analysis and spatial modelling. The availability of GIS stimulates interest in spatial problems, which in turn stimulates interest in spatial analysis. Often the route to spatial analysis begins with using GIS software to integrate a series of different spatial data sets. This is a task which they do rather well, and it would be foolish to reinvent wheels. Once the analyst has carried out the requisite exploratory and/or confirmatory tasks, GIS become invaluable in the display of the results. The problem for the analyst is to find a linkage between the data handling and manipulation functions in a GIS package, and the desired analytical software. Another route is to code one's own functions to achieve the trick – this may take the form of a macro to write the required data to files that can be read by the external software. A filter is also required to read the data back into the GIS software in order to display the results. There is no reason why this should not be a parallel rather than a sequential process. From one's desktop PC or workstation, the GIS software can be run in one window and the analysis takes place in another window, possibly even on another computer. The developments that are taking place to enable advanced exploration and modelling with spatial data, and which take account of the capabilities of computer hardware, are both exciting and stimulating.

Notes

1. Arc/Info is a registered trademark of Environmental Systems Research Institute, Inc., Redlands, CA, USA.
2. The value in the output raster is the average of the eight neighbouring rasters.

4 **Exploring Spatial Data Visually**

4.1 Introduction

It is generally helpful to *look* at a data set before any models are fitted or hypotheses formally tested. This is something that has become much easier over the last decade. Previously, computer graphics were the preserve of a privileged minority of researchers who had access to specialized equipment. However, the advent of window- and mouse-based computer interfaces, coupled with the increasing speed and decreasing cost of computers and the development of appropriate, easy-to-use software, has meant that the production of graphics is now within the reach of virtually all researchers. Interactive graphical displays are the norm on current computing equipment, and most personal computers have sufficient processing power to handle the preprocessing required for many techniques. All of this encourages an exploratory approach to spatial data analysis.

There are a number of reasons for initially 'looking at the data', which can be linked to some basic questions: Are there any variables having unusually high or low values? What distributions do the variables follow? Do observations fall into a number of distinct groups? What associations exist between variables? These questions are informal in nature, and in attempting to answer them, one begins to get a 'feel' for the data.

Initial examination of a data set, by either graphical methods or descriptive statistics, can be a powerful way of moving towards answers to the above questions. In this chapter, a set of useful visual techniques will be outlined. These techniques and this approach in general are often referred to as *exploratory data analysis* (EDA) (Tukey, 1977), and sometimes as *exploratory spatial data analysis* (ESDA) (Unwin and Unwin, 1998) when spatial data are being considered. The ESDA techniques covered in this chapter can be broadly grouped as in Table 4.1. Note that

Table 4.1 **Classification of techniques covered in this chapter**

Univariate	Multivariate
Boxplot	Scatterplot matrix
Histogram	Linked plots
Density estimates	RADVIZ
Maps	Parallel coordinate plots
	Projection pursuit

there is a distinction between univariate and multivariate techniques. This is important, as one explores the *linkages* between variables in different ways to the distribution of individual variables. In particular, some of the multivariate techniques are important for *hypervariate* situations, where there are *more* than three variables under investigation, and some kind of 'projection' of hyperdimensional space to two- or three-dimensional space is neccessary to identify patterns in the data. All of the visualization methods covered here may also be thought of as EDA methods, but clearly not *all* EDA methods involve visualization. Ehrenberg (1982) and Tukey (1977) provide some excellent examples of non-visual EDA.

However, non-visual EDA is not very helpful for identifying spatial patterns – the techniques should be considered in a *spatial* context. Although many of these methods are applicable to non-spatial data, considering the spatial dimension is essential when the data are geographical. Indeed, to the geographer any analysis without a spatial element will be of little value. For example, when a multivariate exploratory method identifies an observation which is unusual in some way, the *location* of the observation will be of interest. If several observations are identified as unusual in some way, are all of these observations in the same geographical region? It is suggested here that issues of this sort are tackled by the use of *linked* graphics – see Section 4.8. Spatial aspects of EDA are also considered in Chapter 5, for examples where observations are *locally* unusual. In this case, spatial arrangement itself plays a role in defining the notion of 'unusual'.

Some other forms of visual data exploration are considered elsewhere in this book. For example, Chapter 6, which discusses the analysis of point patterns, contains a selection of useful visual techniques for summarizing and exploring point data. These include some useful ways of summarizing the location and scale of spatial point distributions, such as the mean centre, and the standard distance. Although these are not intrinsically visual, they provide useful input to some visualization techniques.

4.2 Stem and leaf plots

Table 4.2 shows the percentage rates of heads of households in occupations classed as professional or managerial, by counties in the UK. This will be referred to informally as the PROFMAN index. Surrey has the highest PROFMAN index, and it exceeds the other counties by quite a large amount. However, at the other end of the scale, counties are much more 'bunched together', with three counties more or less tying for bottom place. It is particularly helpful in this case that the table has been listed as two pairs of columns – the values of PROFMAN at the bottom of the first column pair and the top of the second column pair give a good idea of the median of PROFMAN (around 180 households per 1000). Note that when several counties have the same PROFMAN index to three significant digits, the order in which they are tabulated depends on the ordering of their *exact* PROFMAN indices. A stem and leaf plot displays all of the numerical information of Table 4.2 in a

Table 4.2   **Percentage of professional or managerial heads of households by county**

County	PROFMAN	County	PROFMAN
Surrey	31.4	Shropshire	18.0
Buckinghamshire	27.5	Devon	17.9
Hertfordshire	26.8	Leicestershire	17.8
Berkshire	25.2	Suffolk	17.8
West Sussex	24.2	Norfolk	17.7
Oxfordshire	22.0	Cornwall	17.6
Essex	21.5	Northamptonshire	17.4
Hereford and Worcs.	21.3	Lincolnshire	17.1
Warwickshire	20.8	Northumberland	17.0
Gloucestershire	20.7	Staffordshire	16.5
North Yorkshire	20.6	Lancashire	15.8
Cheshire	20.5	Cumbria	15.7
East Sussex	20.1	Derbyshire	15.3
Dorset	20.1	West Yorkshire	15.2
Cambridgeshire	20.1	Nottinghamshire	15.0
Bedfordshire	19.9	Humberside	14.9
Hampshire	19.8	Greater Manchester	14.7
Kent	19.6	Merseyside	13.6
Somerset	19.1	West Midlands	13.5
Avon	19.0	South Yorkshire	12.1
Isle of Wight	18.2	Tyne and Wear	12.0
Greater London	18.2	Durham	12.0
Wiltshire	18.1	Cleveland	12.0

Source: 1981 UK Census of Population

more compact manner. Information is represented to three significant digits. Here the data values are arranged in rows, according to most significant digits (the first fully spanned digit and any partly spanned digit preceding this – this is referred to as the 'stem'). The value of the most significant digits are then used to label the rows. The data in the rows are abbreviated to the third significant digit – 'the leaves'. This is best understood by example. In Table 4.3 (left hand side), the data in Table 4.2 are shown in stem and leaf format.

Note that the leaves are usually shown without spaces between them, so that the row labelled 21 represents the values 21.3 and 21.5. Stem and leaf plots are best shown in a fixed width font, such as Courier, as the length of the leaf display will then be proportional to the number of data items in that row. In this way, a stem and leaf plot gives a near-graphical representation of data distribution while still containing information about each data value within the accuracy of the two digits spanned rule.

A stem and leaf display can be made more compact by combining rows. In the right hand side of Table 4.3 the same information is given, but with only the even-numbered stems. Data relating to odd-numbered stems are shown in italics. The result has removed some of the rough character of the left hand stem and leaf plot. The shape of the distribution, with a large mass of observations between 12% and 20% and a positive tail, becomes apparent.

Table 4.3 **Stem and leaf plot of PROFMAN: left hand is standard form, right hand is compact form**

31	4
30	
29	
28	
27	5
26	8
25	2
24	2
23	
22	0
21	35
20	1115678
19	01689
18	0122
17	01467889
16	5
15	02378
14	79
13	56
12	0001

30	*4*
28	
26	*85*
24	*22*
22	0
20	111567*835*
18	012*201689*
16	*5*01467889
14	*7902378*
12	0001*56*

4.3 Boxplots

A useful set of descriptive statistics is the *five-number summary*. This uses *order-based statistics*, the median, quartiles and extreme values. The median of a variable is simply the middle value if the variable is tabulated in ascending order and n is odd. If n is even it is the midpoint of the two middle values. Thus, it is effectively the 'half-way point' of the distribution. Again, this represents a 'typical value' of the variable. The quartiles are defined similarly, being the points one-quarter and three-quarters along the sorted list of a variable. A useful measure of spread is the *interquartile range*, the difference between the first and third quartiles. The extreme values are just the largest and smallest values of the variable in the data set. The five-number summary is the list (minimum value, first quartile, median, third quartile, maximum). Listing these numbers gives a good impression of the location, spread and extreme values of a data set. The values for PROFMAN are listed in Table 4.4. Again, it is possible to see the skewness of the distribution, by noting the position of the median value in relation to the range.

Order-based statistics are less likely to be affected by *outliers* than the mean or standard deviation. An outlier is defined to be an observation with an unusually large or small value of some variable. The midpoint (in order terms) of a set of

Table 4.4 **Five-number summary for PROFMAN**

Minimum	First quartile	Median	Third quartile	Maximum
12.0	15.8	18.1	20.3	31.4

numbers will not be affected by the values of points at the extremes, and neither will the first or third quartiles. It is sometimes interesting to compare the mean and the median; in general the mean is 'pulled' towards the longest tail of the distribution if there is some degree of skewness. For example, the mean of PROFMAN is 18.5%, but the median is 18.1%.

The boxplot (or box and whisker plot) is the graphical equivalent of the five-number summary. In its crudest form, a box and whisker plot is set out as in Figure 4.1. The two *whiskers* extend to the minimum and maximum values of the sample, and the *box* extends from the lower to the upper quartile. A vertical stripe on the box indicates the location of the median. The skewness of the distribution is readily apparent.

A more refined approach to boxplots attempts to show individual outlying values. In this case, the whiskers do not always extend to the extreme values. For example, the left hand whisker could extend to max(x_{min}, $Q_2 - 1.5(Q_3 - Q_1)$) and the right hand whisker could extend to min(x_{max}, $Q_2 - 1.5(Q_1 - Q_3)$), where Q_1 is the first quartile, Q_2 is the median and Q_3 is the third quartile. If the distribution is fairly compact, the whiskers will in fact be the extreme values. However, if there are values of the variable differing from the median by more than the interquartile range times one and a half in either direction, the whiskers will not extend as far as these values. In this case, the values should be marked individually. For the PROFMAN variable, this is shown in Figure 4.2. The underlying message here is that the distribution is skewed, but that much of the skewness is due to four outlying observations in the higher end of the distribution.

One very useful property of box and whisker plots is that they are relatively flat, so that several may be stacked vertically. This is helpful as a way of comparing the distributions of several different variables measured on the same scale – or several variables which have been standardized to have zero mean and unit variance.

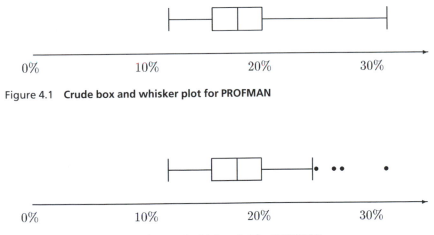

Figure 4.1 **Crude box and whisker plot for PROFMAN**

Figure 4.2 **Outlier showing box and whisker plot for PROFMAN**

4.4 Histograms

Histograms are an alternative way of representing the distribution of a variable. The sample range of a variable is divided into several intervals (referred to as *bins*) and the number of variables in each bin is counted. Usually, each bin has the same width. Then a bar graph is drawn, with the height of each bar being proportional to the number of values of the variable occurring in each bin. A histogram of the PROFMAN variable discussed above is shown in Figure 4.3.

A variation on a histogram is a *frequency polygon*, shown in Figure 4.4 for the PROFMAN variable. Here, the midpoints of the bars are joined together with lines. Both the histogram and the frequency polygon may be thought of as (scaled) approximations to the probability density function of x. The histogram approximates by assuming the probability density is constant within each bin, whilst the frequency polygon uses a piecewise linear approximation.

An important issue with histograms (and frequency polygons) is the bin width. Very large bin widths tend to obscure detail, while very small bin widths tend to produce graphs which are too rough. For examples of the effects of this (with the PROFMAN data) see Brunsdon (1995a). One reasonable 'automatic'[1] choice of bin width is the conservative approach of Terrell and Scott (1985). This approach finds a *maximal smoothing* histogram. If one considers an 'optimal' sample histogram in the sense that it gives the best mean squared error of the true distribution, then an optimal bin width can be found. The maximal smoothing bin width is the largest

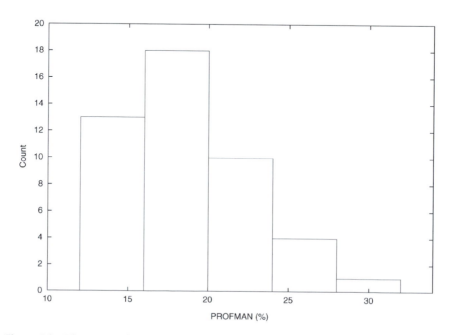

Figure 4.3 **Histogram of PROFMAN**

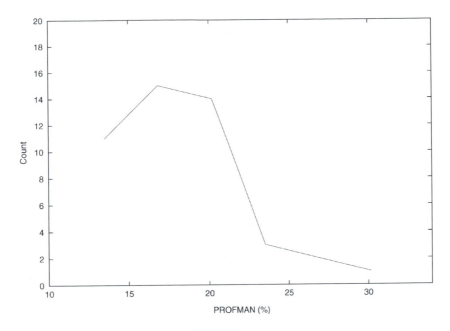

Figure 4.4   **Frequency polygon of PROFMAN**

bin width one could obtain for *any* possible distribution function. Terrell and Scott show that this requires about $\sqrt[3]{2n}$ bins, where *n* is the number of observations. The approach is conservative in the sense that it is probably oversmoothing (unless the true probability is actually the maximal optimal smoothing case), so that any modes of the observed histogram are likely to correspond to real modes, and not likely to be artefacts of random sampling. It is therefore a reasonably 'foolproof' choice. In the same paper, a similar rule is given for frequency polygons. In this case maximal smoothing occurs with about $\sqrt[5]{73.5n}$ bins. The numbers of bins in Figures 4.3 and 4.4 were chosen using these formulae.

4.5   Density estimates

The frequency polygon considered in the last section may be thought of as a way of estimating the probability density function for a variable *x*. A slightly more sophisticated approach is the *kernel density estimate* (Silverman, 1986; Brunsdon, 1995b). This approach attempts to provide a *smooth* estimate of probability density. To achieve this, a small hump (the *kernel*) is centred on each x_i. These kernels are then averaged to obtain an estimate of the probability density function of *x*. The kernel is itself a probability distribution function, typically unimodal and symmetrical, often taking the form

$$K(x) = \frac{1}{h} g\left(\frac{x - x_i}{h}\right) \tag{4.1}$$

where g is a probability distribution function with mean zero and variance one. Thus, the density estimate of $f(x)$, say $\hat{f}(x)$, is given by

$$\hat{f}(x) = \sum_{i=1}^{n} \frac{1}{nh} g\left(\frac{x - x_i}{h}\right) \tag{4.2}$$

The term h (called the *bandwidth*) controls the spread of K: very large h gives a virtually flat kernel, and h close to zero gives a sharp spike centred on x_i. Thus, h plays much the same role as the bin width in a histogram: too large a value smooths out detail, and too small a value causes a spikey estimate. Terrell (1990) gives a maximal smoothing rule for choosing h similar in nature to the rules for choosing histogram bin widths discussed in the last section. This is given by

$$h_{\max} \approx s \sqrt[5]{\frac{243 \int g(x)^2 \, dx}{35n}} \tag{4.3}$$

where $h_{\max}$ is the maximal smoothing bandwidth, and s is the sample standard deviation. Note that this will depend on the choice of $g(x)$. For a normal g, we have

$$h_{\max} \approx \frac{1.144s}{\sqrt[5]{n}} \tag{4.4}$$

Using a normal kernel, the density estimate for PROFMAN is shown in Figure 4.5. Kernel density estimates for *spatial* data are considered in Chapter 6, where point pattern analysis is considered.

4.6   Maps

In exploratory spatial data analysis the map has an important role to play. An important factor when creating univariate maps is the avoidance of spurious detail. For example, consider a standard choropleth map, such as the one in Figure 4.6. This shows crime rates (per 10 000 households) for zones in a study area. The boundaries of the zones are indicated to a great deal of precision, but the levels of crime rates – which it is claimed are the subject matter of the map – are crudely classed into five discrete categories! Worse still, the outlining of ward boundaries with black curves draws even more attention to the shapes of wards, and away from spatial patterns in crime rates.

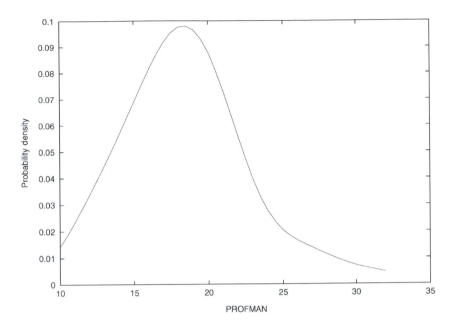

Figure 4.5   **Kernel density estimate for PROFMAN**

This type of map appears quite frequently, for a number of possible reasons. One is that they are relatively easy to produce using 'standard' desktop mapping and GIS software packages. Another is that people have become used to them. However, there are a number of alternative ways to produce maps from the same geographical information as that used in Figure 4.6. First, the ward boundaries need not be included in the map. One could simply draw the map by shading the ward polygons, excluding the boundaries, as exemplified in Chapter 3. This is a particularly helpful suggestion when there is a great deal of variation in the size of wards, since in this case very small wards can have a maximum width not much larger than the width of the black ward boundary lines in Figure 4.6 – and the inclusion of these lines can then tend to obscure spatial pattern.

This still leaves the problem of crude classification. One way of overcoming this is to use *unclassed* choropleth maps. Here the intensity of shading of zonal polygons varies continuously with the variable of interest. Support for this approach may be found in Tobler (1973a, p. 263):

The main argument in favour of class intervals is that their use enhances readability. This at least is the assertion . . . if the assertion is in fact valid why then is a group of greys into classes not also (e.g. in addition to spatial filtering) used to enhance areal photographs, or television?

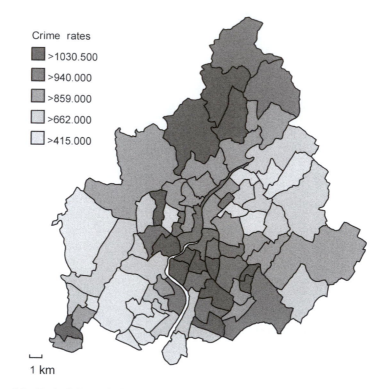

Crime rates
■ >1030.500
■ >940.000
■ >859.000
□ >662.000
□ >415.000

1 km

Figure 4.6 **Typical choropleth map**

Thus, an alternative choropleth representation of crime rates is given in Figure 4.7, which uses no shade classes and adds no solid ward boundaries.

Despite the above changes in format, the map in Figure 4.7 still highlights ward boundaries to some extent. The only way to avoid this is to use some kind of map format that illustrates ward-based data without actually drawing ward polygons. One such possibility is the *random dot map*. Here, for each ward, a number of dots proportional to the variable to be mapped are placed at random points within the ward. If the ward boundaries themselves are not drawn then a map of varying dot intensity is produced, with the dot intensity varying in proportion to the variable of interest. This is illustrated in Figure 4.8.

One could argue that the last approach is somewhat dishonest, as it tries to hide the fact that the map was produced using aggregated data. It could be said that making ward boundaries evident in some reasonably unobtrusive way on the map provides a warning sign. The *modifiable areal unit problem* (Openshaw, 1984; Fotheringham and Wong, 1991), discussed in Chapter 10, implies that spatially aggregated data contain a higher degree of uncertainty than the individual components undergoing aggregation, and that some observed patterns could well be artefacts of the aggregation process. Thus, some indication that aggregated data are being displayed does at least suggest that some of these problems may be present.

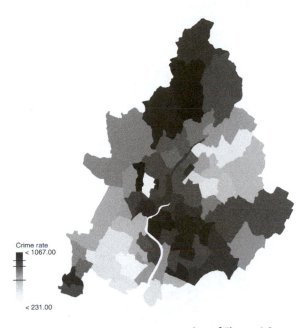

Figure 4.7   **Unclassed choropleth map representation of Figure 4.6**

This would suggest that a map of the type of Figure 4.7 might be the most suitable, or that if a map of the kind in Figure 4.8 is used, then at some place on the map it should be clearly labelled that aggregate data are being used, and the level of aggregation stated.

Finally, when considering maps in human geography, the concept of the *cartogram* may be helpful. Essentially, a cartogram can be thought of as a map projection, such that the areal density of some attribute is uniform across the map. Often the attribute used to define the cartogram is population distribution. This allows map patterns in and around city centres (where population is usually highest) to be seen more clearly. Any of the kinds of map discussed above can be displayed in cartogram format – one simply applies the cartogram transformation to all of the data being mapped. Two useful texts on cartograms are Dorling (1991) and Tobler (1973b).

4.7   The scatterplot matrix

The previous techniques have all considered exploration of an individual variable. However, in most studies it is as important to consider *relationships* between variables. The remaining material in this section addresses this objective. In order to decide how useful a graphical representation is, one needs to consider the kind of feature in data that one wishes to detect. Three common possibilities are *clusters*, *outliers* and *trends*. Clusters are distinct groupings in the data points, usually

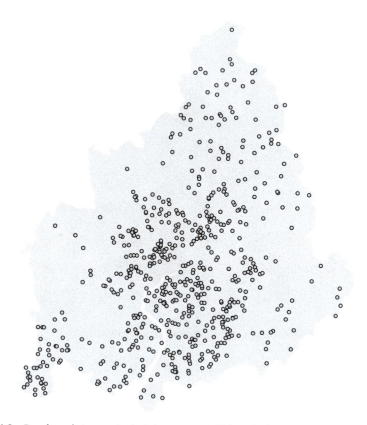

Figure 4.8   **Random dot map. Each dot represents 10 burglaries per 10 000 households**

corresponding to multimodality in the underlying probability distribution for the data. Outliers are one-off cases that have unusual combinations of observed values, when compared with the remainder of the sample. Geographical trends are fairly self-explanatory. To some extent, each of the following techniques could be used to identify any of these kinds of feature.

When investigating the relationship between a pair of variables, a *scatterplot* is a useful tool. This simply plots each observation as a two-dimensional point on a graph whose axes are the two variables of interest. This is a useful way of identifying trends, clusters or outliers, as long as they are not so subtle as to be undetectable in two dimensions. This is an improvement on using univariate methods. For example, it is possible to find a two-dimensional outlier which does not take unusual values in either of its one-dimensional components – see Figure 4.9.

A *scatterplot matrix* simply takes all of the possible pairs of variables in a data set of *m* variables, and arranges the corresponding scatterplots into a rectangular array of plots. Each scatterplot in a row has the *y* variable in common, and each

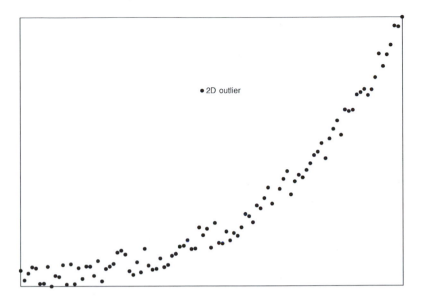

Figure 4.9   **A two-dimensional outlier**

scatterplot in a column has the *x* variable in common. This allows *all* two-way patterns in a data set to be seen simultaneously. Typically a scatterplot matrix gives less detail on the axes than a single scatterplot. This avoids clutter if *m* is relatively large. A typical scatterplot display for the limiting long-term illness data described in Chapter 5 is shown in Figure 4.10. This data set consists of the variables described in Table 4.5 for census wards in northern England. The scatterplot matrix was created using the XLispStat package (Tierney, 1990). Inspecting this matrix identifies a number of outlying points – for example, there is one observation with a particularly unusual combination of the LLTI and SPF variables, represented by a point on the far left of the SPF vs. LLTI panel in the matrix. Certain trends also become apparent – for example, a positive connection between LLTI and UNEMP can be seen. It should be noted, however, that when scatterplot matrices are used to depict highly multivariate data (say seven or more dimensions) it is debatable how much detail can be seen. Another limitation is that they are only capable of showing *pairwise* relationships between variables. The following sections discuss techniques which address this limitation.

4.8   Linked plots

Looking at scatterplot matrices, one can check for outliers (or clusters) in various pairs of variables. Suppose we have four variables, say $x_1 \ldots x_4$, and we see one outlier in the plot for x_1 against x_2, and another in the plot for x_3 against x_4. Here, it would be useful to discover whether the pair of outliers seen in these two plots

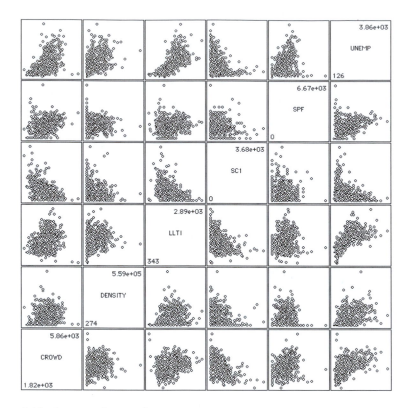

Figure 4.10   **A scatterplot matrix**

Table 4.5   **Variables in the LLTI data set**

VARIABLE	DESCRIPTION
LLTI	The percentage of persons in households in each ward where a member of the household has some limiting long-term illness. This is the response variable. Note that to control for different age profiles in areas, this is only computed for 45–65 year olds – an age category that is perhaps most likely to suffer LLTI as a result of working in the extractive industries.
CROWD	This is the proportion of households in each census ward having an average of more than one person per room. This is an attempt to measure the level of cramped housing conditions in each ward.
DENSITY	This is the housing density of each ward, measured in millions per square kilometre. This is intended to measure 'rurality' of areas. Note the differences between this and the previous variable – a remote village with poor housing conditions may well score low in this variable, but high in the previous one.
UNEMP	The proportion of male unemployment in an area. This is generally regarded as a measure of economic well-being for an area.
SC1	The proportion of heads of households whose jobs are classed in social class I in the Census. These are professional and managerial occupations. Whilst the previous variable measures general well-being, this measures affluence.
SPF	The proportion of single-parent families in an area. This is an attempt to measure the nature of household composition in areas.

correspond to the same case. This allows one to distinguish between the following two situations: one single case which is a four-dimensional outlier, or two cases each of which is a two-dimensional outlier. *Linking* the plots in a scatterplot matrix allows this to be investigated. When a point is selected on one plot (by pointing and clicking with the mouse on a computer) this causes the point to be highlighted, by an enlargement or a change of colour or both. At the same time, all points corresponding to the same case on the other plots are also highlighted. Thus, it is possible to check whether the outlying point on one of the scatterplots mentioned in the example corresponds to the other.

Linked plots as described above allow one to investigate higher levels of interaction between groups of variables. These plots may also be used to check relationships between trends or clusters. For example, by highlighting every point in a cluster appearing on one plot it is possible to see whether this corresponds to clusters in any other plots.

A similar approach can be used to link scatterplots (or scatterplot matrices) to maps. Here, if each case in a data set corresponds to a geographical zone, line or point, then selecting a point in a scatterplot could highlight the corresponding geographical entity on a map. For example, it would be possible to identify the geographical locations of the outliers discussed in Section 4.7 using this approach. Similarly, selecting the geographical entity on the map would cause the corresponding point to be highlighted on the plot. An example of this is illustrated in Figure 4.11. Here, the points in the upper right hand part of the scatterplot in the left hand window are selected with the mouse. The corresponding zones are highlighted on the map (lighter shading) in the right hand window, showing a 'c'-shaped formation, with one geographically isolated zone to the south-west of the map. Equivalently, we could select the outlying points in the scatterplot and identify their geographical locations on the linked map.

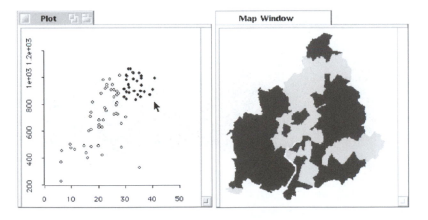

Figure 4.11 **A linked map and scatterplot**

4.9 Parallel coordinate plots

One shortcoming of scatterplot matrices is that they only show interactions between *pairs* of variables. If there are m variables, one may wish to consider anything up to m-way interactions between variables. In this section an approach allowing this is described. Suppose, as before, that there are m variables in a data set. In the *parallel coordinates* approach, a point in m-dimensional space is represented as a series of $m - 1$ line segments (Inselberg, 1985) in two-dimensional space. Thus, if the original data observation is written as $(x_1, x_2, \ldots, x_m)$ then, its parallel coordinate representation is the $m - 1$ line segments connecting the points $(1, x_1)$, $(2, x_2) \ldots (m, x_m)$. Each set of line segments could be thought of as a 'profile' of a given case. The shape of the segments conveys information about the levels of the m variables. Typically, continuous variables will be standardized before a parallel coordinate plot is drawn.

To view an entire m-dimensional data set one simply plots *all* such profiles on the same graph. This is illustrated in Figure 4.12, in which the limiting long-term illness data set is plotted. For large data sets, the appearance of such a plot appears confusing, but can be used to highlight outliers. However, the real strength of the technique can be seen when subsets of the data are selected – usually on the basis of one particular variable. To see this, consider Figure 4.13. Here, the subset of the data in the lowest decile of the variable LLTI is shown in black, and the remainder of the data set in grey. Looking at the relative locations of the black and grey lines shows the distribution of the data values in the subset in relation to the entire data set. Obviously, all of the black lines pass through the lowest section of the LLTI

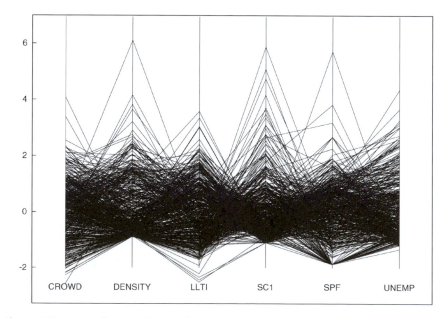

Figure 4.12 **A parallel coordinates plot**

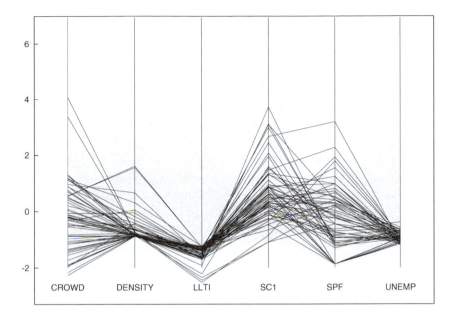

Figure 4.13   Parallel coordinates plot, lowest decile of LLTI highlighted

axis. However, looking at the locations of the black lines on the other axes shows whether the low values of this variable tend to be accompanied by any notable distributional patterns in the other variables. From the plot, it may be seen that often there are also low values of DENSITY and UNEMP.

Parallel plots may also be used to detect outliers in two dimensions. Again looking at Figure 4.13, there are a few cases in the subset where DENSITY is unusually high, *given* the low value of LLTI. It is also apparent that this phenomenon does not occur with the variable UNEMP. This technique can also be used, at least sometimes, to detect three-dimensional outliers. For example, the line joining a high(ish) value of SC1 to a similar value of SPF is unusual, first in a two-dimensional sense because it appears that the two variables *both* have high values, and secondly in three dimensions – we can also see that this line associated with the lowest decile of LLTI.

As with scatterplot matrices, and the methods of RADVIZ and projection pursuit discussed below, parallel coordinates plots may also be used to investigate spatial pattern. A powerful and interactive way to achieve this is to use the 'linking' approach discussed in Section 4.8. Highlighted parallel lines (as seen in Figure 4.13) are linked to zones (or points) on a map, so that the locations of highlighted cases can be investigated. As before, it is also possible to consider a linkage in the opposite direction. That is, by selecting an observation region in the study area one can see the corresponding parallel coordinates highlighted. This allows the exploration of the distribution of several variables associated with a geographical subset of the data to be considered in relation to the distribution of the same variables for the *whole* study.

Outliers detected in terms of the lines connecting pairs of axes in the parallel system pose an interesting problem. Although the method provides a striking image of outliers between two variables, it only works if the two variables have neighbouring parallel axes. For m variables, there are only $(m-1)$ such neighbours possible, but there are $m(m-1)/2$ possible variable pairs. Thus, $(m-1)(m-2)/2$ pairs cannot be displayed. The problem is similar to the ordering problem in RADVIZ discussed in Section 4.10 – the patterns that parallel coordinates plots yield depend on the ordering of the axes. For a parallel coordinates plot, there are $m!$ possible orderings, although if we assume that reversing the order of the axes generates equivalent patterns, this leaves $m!/2$ possibilities. A number of ways of automatically choosing an ordering could be used, usually based on maximizing some criterion measuring the 'strength of pattern' displayed. Alternatively, ordering could be controlled interactively by the analyst. This second approach is perhaps more in the spirit of exploratory data analysis.

4.10 RADVIZ

Like the previous approach, the RADVIZ method (Hoffman et al., 1997; Ankerst et al., 1996) maps a set of m-dimensional points onto two-dimensional space. To explain the approach, it is helpful to imagine a physical situation. Suppose m points are arranged to be equally spaced around the circumference of the unit circle. Call these points S_1 to S_m. Now suppose a set of m springs is fixed at one end to each of these points, and that all of the springs are attached to the other end to a puck, as in Figure 4.14. Finally, assume the stiffness constant (in terms of Hooke's law) of the jth string is x_{ij} for one of the data points i. If the puck is released and allowed to reach an equilibrium position, the coordinates of this position, $(u_i, v_i)^T$ say, are the projection in two-dimensional space of the point $(x_{i1}, \ldots, x_{im})^T$ in m-dimensional space. Thus, if $(u_i, v_i)^T$ is computed for $i = 1 \ldots n$, and these points are plotted, a visualization of the m-dimensional data set in two dimensions is achieved.

To discover more about the projection from R^m to R^2, consider the forces acting on the puck. For a given spring, the force acting on the puck is the product of the vector spring extension and the scalar stiffness constant. The resultant force acting on the puck for all m springs will be the sum of these individual forces. When the puck is in equilibrium there are no resultant forces acting on it and this sum will be zero. Denoting the position vectors of S_1 to S_m by $\mathbf{S}_1$ to $\mathbf{S}_m$, and putting $\mathbf{u}_i = (u_i, v_i)^T$, we have

$$\sum_{j=1,m} (\mathbf{S}_j - \mathbf{u}_i)x_{ij} = 0 \qquad (4.5)$$

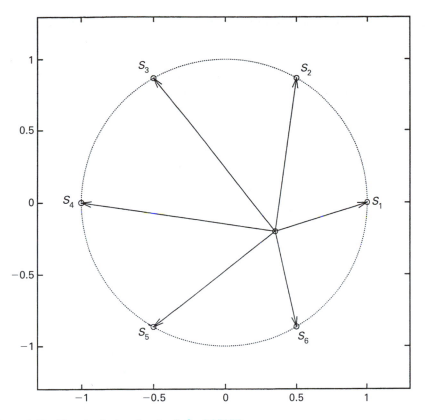

Figure 4.14 **The physical system basis for RADVIZ**

which may be solved for $\mathbf{u}_i$ by

$$\mathbf{u}_i = \sum_{j=1,m} w_{ij}\mathbf{S}_j \qquad (4.6)$$

where

$$w_{ij} = \left(\sum_{j=1,m} x_{ij}\right)^{-1} x_{ij} \qquad (4.7)$$

Thus, for each case i, $\mathbf{u}_i$ is simply a weighted mean of the $\mathbf{S}_j$ whose weights are the m variables for case i normalized to sum to one. Note that this normalization operation makes the mapping from R^m to R^2 non-linear.

Viewing the projection in this explicit form allows several of its properties to be deduced. First, assuming that the x_{ij} values are all non-negative, each $\mathbf{u}_i$ lies within the convex hull of the points $\mathbf{S}_1$ to $\mathbf{S}_m$. Owing to the regular spacing of these points, this convex hull will be an m-sided regular polygon. Note that if some of the x_{ij} values are negative, this property need not hold, but that often each variable is rescaled to avoid negative values. Two typical methods of doing this are the local metric (L-metric) rescaling, in which the minimum and maximum values of x_{ij} for each j are respectively mapped onto zero and one respectively,

$$x_{ij}^{\mathrm{L}} = \frac{x_{ij} - \min_j(x_{ij})}{\max_j(x_{ij}) - \min_j(x_{ij})} \tag{4.8}$$

and the global metric (G-metric), in which the rescaling is applied to the data set as a whole, rather than on a variable-by-variable basis:

$$x_{ij}^{\mathrm{G}} = \frac{x_{ij} - \min_{ij}(x_{ij})}{\max_{ij}(x_{ij}) - \min_{ij}(x_{ij})} \tag{4.9}$$

In each case, the rescaled x_{ij} values will all lie in the interval [0, 1].

The weighted centroid interpretation of the projection also allows some other properties to become apparent. If, for a given i, the values of x_{ij} are constant, $\mathbf{u}_i$ will be the zero vector. This is a rather strange property, since it implies that observations in which all variables take on a very high constant value (once rescaled) will be projected onto the same point as observations in which all variables take on a very low constant value. More generally, any point plotted on the RADVIZ circle does not correspond to a *unique* set of x variables.

For general data sets this property could lead to difficulties in interpreting the plots, but it is particularly useful when considering *compositional* geographical data. Suppose the population of a geographical region is classified into m categories, for example those aged under 18, those aged 18 to 65 and those aged over 65. Another example would be voting data for an electoral ward – here the categories are the parties for which each constituent voted. A compositional data set is one in which each of the variables represents the proportion (or percentage) of each category for each area. Numerically, the most notable property of such data is that for each case the variables sum to 1 (or 100 if percentages are used). This constraint suggests that the only way that all m variables can be equal is when they all take the value $1/m$. Note, however, that even in this case, the fact that $\mathbf{u}_i = 0$ does not imply that all proportions are equal. For example, if a pair of variables are represented by diametrically opposite points on the circle, and the proportions are 0.5 in each of these, then this will also give $\mathbf{u}_i = 0$. Another aid to interpretation for compositional data is that if an area consists entirely of one category then the corresponding variable will take the value 1, while the others will take the value zero. This implies that $\mathbf{u}_i$ will lie on the vertex of the regular m-sided polygon corresponding to that category.

In the case $m = 3$ for compositional data the RADVIZ procedure produces the compositional triangular plot or tripartite plot used for various purposes – notably by Dorling (1991) to illustrate voting patterns in the UK in a three-party system. If we were to extend the analysis to look at more than three parties then RADVIZ provides a natural extension of this concept. The main difficulty when moving beyond $m = 3$ for compositional data is that points on a RADVIZ plot no longer correspond *uniquely* to $(x_{i1} \ldots x_{ij})$ for a given case – more than one composition can project onto the same $\mathbf{u}_i$, as discussed above.

A RADVIZ projection for the limiting long-term illness data is shown in Figure 4.15. The result shows a circular cluster of points and identifies outliers around this. In particular, three points lie to the right of the main cluster, towards the CROWD and UNEMP fixed points. This suggests that there are three places with particularly large values of these two variables, in relation to the other variables.

A further pattern in Figure 4.15 is that a lower-density 'cloud' of points lies to the left of a central cluster. This seems to be biased towards the direction of SC1 on the RADVIZ frame suggesting that the areas associated with this cloud tend to have relatively high levels of 'social class I' inhabitants. Further insight can be obtained by linking the RADVIZ plot to a map as in the screenshot in Figure 4.16. This is achieved by linking the plot to a map of the study region and highlighting the points in the cloud of the RADVIZ plot. The geographical locations are then highlighted in the map window. From this it may be seen that the cloud links mostly to the more rural areas in the study region.

It is also interesting to note that for a given set of variables, there are several

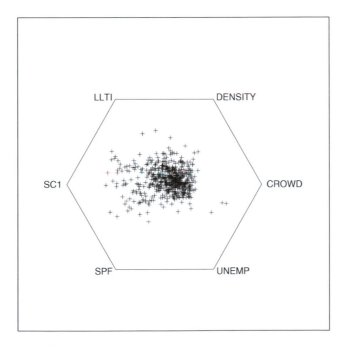

Figure 4.15 **A RADVIZ projection of the census data**

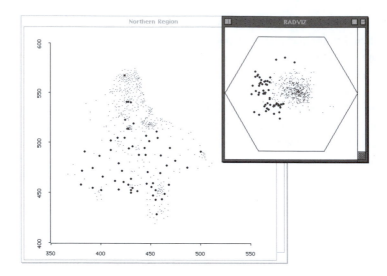

Figure 4.16   **Linking RADVIZ to a map**

possible RADVIZ projections, since the initial m variables could be assigned to $S_1 \ldots S_m$ in $m!$ different ways. If we are mostly interested in identifying clusters and outliers, a number of possible projections will be essentially equivalent, since they will be identical up to a rotation or a mirror image. To see how many non-trivially different permutations there are, we need first to note that any permutation can be rotated m ways (i.e. rotation through $360°/m$ by $2(360°/m)$ and so on up to $(m - 1)(360°/m)$, and of course the identity rotation through zero degrees), and so we need to divide the $m!$ by m. We then note that any permutation can be reflected in two ways (i.e. mirror imaged or left alone) so the figure of $(m - 1)!$ must be halved. Thus, if m is the number of variables, there are effectively $(m - 1)!/2$ possible RADVIZ projections.

One way of deciding which of these should be used is to use an index, in a similar manner to projection pursuit in the previous section. In fact, the same indices could be used – for example, maximizing the variance of the $\mathbf{u}_i$. In this case, the optimization is a discrete search over a finite number of possibilities, rather than a continuous multivariate optimization problem as in projection pursuit. It should be noted that the number of options to be searched increases very rapidly with m – worse than m^2 – and this has implications for computation, namely a similar situation to the axis ordering in the parallel coordinates approach. Once again, some interactive control of the projection may be a practical alternative.

4.11   Projection pursuit

Suppose as before that for a set of cases m variables are recorded. Then each case can be thought of as a point in m-dimensional space. Unfortunately, unless $m \leqslant 3$

it is not possible to view these points directly. However, it is possible to *project* an *m*-dimensional set of points onto a two-dimensional plane, or a three-dimensional volume. Here we will restrict the problem to projections onto two-dimensional planes. To visualize the concept of projection, Figures 4.17 and 4.18 should be considered. In both figures a rectangular 'screen' is shown, either above or to the right of a set of three-dimensional data. Imagine a very bright light on the other side of the data points. The shadows thrown on the screen from the data points are the projection. The dotted lines in the diagram link the data points to their projected images.

In Figure 4.17 the data points are projected onto a plane to the right. Here the projected image shows two distinct clusters of points. In Figure 4.18 the same data points are projected onto a plane above. Here the projected image shows only a single cluster of points. Obviously in this case the projection is from R^3 to R^2, but similar (and sometimes more complex) phenomena occur when the projection is from R^m dimensions and $m > 3$.

The above example demonstrates that different projections of the same data set can reveal different aspects of the data structure – indeed some projections can fail to reveal any structure. There are in fact an infinite number of possible projections to choose from, so which one should be used? This problem was initially posed by Kruskal (1969), and the term *projection pursuit*, coined by Friedman and Tukey (1974), refers to resolving this kind of problem.

The initial problem is to decide what kind of feature one wishes to detect. When this decision is made, one attempts to measure the degree to which this feature is exhibited in a given projection. For example, Jones and Sibson (1987) and Huber

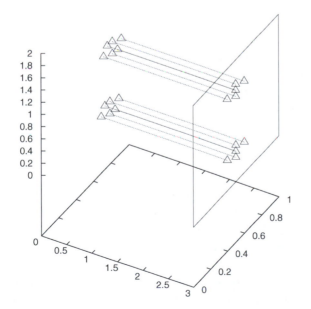

Figure 4.17   **Example of point projection (1)**

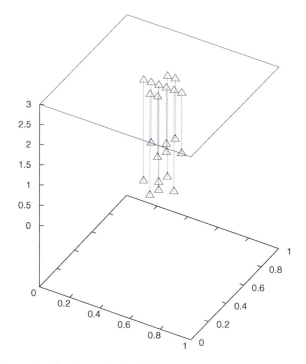

Figure 4.18   **Example of point projection (2)**

(1985) consider departures from a Gaussian distribution shape for the projected points to be of interest. This viewpoint can be justified by considering the findings of Diaconis and Freedman (1984), which suggest that randomly chosen projections of sets of points from high-dimensional spaces to two or three dimensions tend to give sets of projected points that appear to follow a Gaussian distribution. Finding any projection that is an exception to this can be regarded as finding an interesting projection. Another approach suggested by Cook et al. (1993) finds projections that emphasize central 'holes' in the distribution of projected points, and also projections that emphasize skewness. In each case, a characteristic is associated with the 'interestingness' of any projection, and the specific projection optimizing this characteristic is sought.

To illustrate this idea an example is now given. Suppose one equates 'interestingness' with clusters in the projected data. A common test statistic for clustering in two-dimensional data is the *mean nearest-neighbour distance* (MNND). Lower values of this statistic indicate greater clustering. In this case, the MNND is the measure of clustering. The 'best' projection choice problem can be thought of as an optimization problem in which the projection must be chosen to optimize this measure.

In Figure 4.19 the result of applying this technique to the limiting long-term illness data is shown. Owing to the nature of the index function, the rotation of the point pattern obtained is arbitrary, so there is no clear interpretation of the

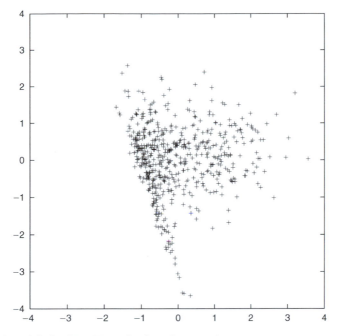

Figure 4.19   **Minimized MNND projection of census data**

individual axes in the plot. No obvious clusters exist in the plot, suggesting perhaps that the data are not multimodal in any way detectable by projecting onto a two-dimensional plane. However, some features are very clear, most notably a 'spur' in the lower part of the plot. By linking this plot to a map in the manner suggested in Sections 4.8 and 4.10 one can check to see whether this spur is a regional effect, or is spatially diverse in nature. For example, one can check whether the spur in the minimizing MNND projection pursuit corresponds to any particular geographical pattern, as in the screenshot in Figure 4.20. Here, as with the RADVIZ example, the points identified as unusual in the projection plot correspond to areas in rural parts of the southern region of the study area.

Another useful interactive approach is *slicing* as described in Tierney (1990). In this case, points in a scatterplot are selected according to the value of an auxiliary variable. This value is controlled by a slider button, as in the second screenshot, Figure 4.21. The value shown in the slider is the central point of a decile 'slice' of the data, based on the values of the variable LLTI – clearly any other variable could also be used. Moving the slider causes the highlighted points in the scatterplot to change, so one can see which regions of the projection correspond to high and low values of the slicing variable. This method helps to interpret the patterns seen in projection-based methods such as those considered in this section.

An important concept related to projection pursuit is the *grand tour* (Asimov, 1985). One shortcoming of the former technique is that it is necessary to define 'interestingness' before viewing the data. If the data happen to have some pro-

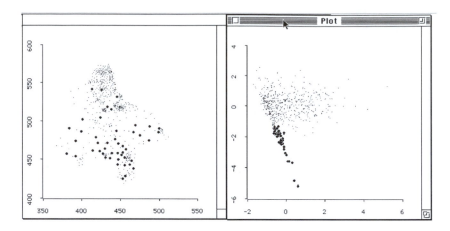

Figure 4.20 **A map-linking screenshot**

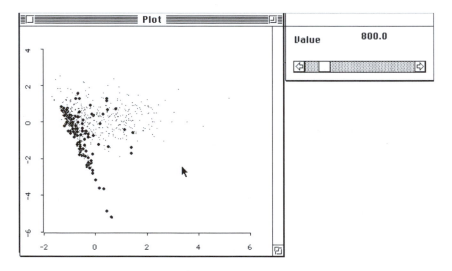

Figure 4.21 **A slicing screenshot**

jections which are interesting in some other way than this formal definition, it is possible that these projections will not be discovered. The grand tour addresses this by attempting to display all projections of the data in one or two dimensions. To do this, an animated display of the projected points is used. As the plane (or line) of projection moves continuously in the *m*-dimensional space of the data, the locations of the projected data points on the plane (or line) also move continuously. When the path of motion for the plane of projection eventually traces through *all* possible planes, then the animated point display will have shown all possible projected sets of points. Grand tour software exists in a number of packages, for example XGobi (Swayne et al., 1991) and Lisp-Stat (Tierney, 1990) both of which are in the public domain. In a sense, the grand tour complements projection pursuit: projection pursuit shows a snapshot of a single 'interesting' projection, whilst the grand tour shows a movie of all possible projections. The drawback with the latter option is that one could sit through a very long movie of uninteresting projections before an interesting one is seen!

To address this problem, the notion of *guided* grand tours is useful. Rather than considering every possible projection, a path concentrating on interesting projections is used. Cook et al. (1995) combine this idea with the index of 'interestingness' used in projection pursuit. Typically, the projection pursuit index is optimized using an iterative procedure, so that a series of successively more and more interesting projections is produced, terminating at a local optimum of 'interestingness'. If a movie is created using this sequence of projections, then a guided grand tour is created. Noting that the optima found in this way are not always global, once an optimum has been achieved a random projection is selected and the iterative process restarted. With this method it has been found that interesting non-optimal projections can be discovered, and that the dynamic can provide context between different projections (Cook et al., 1995). For example, the path of a point which is outlying in one projection can be observed to see whether the same point is an outlier in another projection.

Finally, it has also been noted that in contrast to either of these *automatic* approaches to finding projections, there are occasions when *manual* control of the projection can be very useful (Cook and Buja, 1997). Here, one uses a set of controls to move the plane of projection, and detects interesting point arrangements by eye. Note that this facility, as well as the guided grand tour, are available in XGobi. A strong geographical application of these techniques is provided by Cook et al. (1998) who link XGobi to the C2 virtual reality environment[2] in order to explore patterns in a number of environmental variables recorded on 501 samples of streams in several mid-Atlantic states in the USA.

4.12 Summary

The most important tool in the visualization of spatial data is the map. Clearly maps enable the visualization of a simple, univariate spatial pattern, as is evidenced by their abundant use throughout several of the chapters in this book. However, current spatial data exploration techniques offer more than static maps. In particular, linked maps offer greater opportunities to investigate the spatial aspects of relationships between large numbers of variables when used in conjunction with multivariate data visualization tools such as RADVIZ, projection pursuit or parallel coordinates plots. Such linkages offer not only the ability to investigate outliers and trends in multidimensional data, but also the ability to consider this in a spatial context. Questions such as 'where are the outliers located?', and 'does a multi-dimensional cluster of variable values correspond to a geographical cluster?', may be answered in an informal framework using the techniques outlined in this chapter.

Exploratory spatial data analysis may also be used to examine, for 'spatial outliers' – that is for observations that are very different from others in their *locality* although not necessarily outstanding in the data set as a whole. This concept is discussed further in Chapter 5.

Notes

1. 'Automatic' is not necessarily optimal!
2. They also link XGobi to ArcView with less striking results.

5 Local Analysis

5.1 Introduction

Traditionally, spatial models and methods of spatial analysis have been applied at a 'global' level, meaning that one set of results is generated from the analysis and these results, representing one set of relationships, are assumed to apply equally across the study region. Essentially, what is being undertaken in a global analysis, but is rarely acknowledged, is the generation of an 'average' set of results from the data. If the relationships being examined vary across the study region, the global results will have limited application to any particular part of that region and may not, in fact, represent the actual situation in any part of it. In a global analysis, we typically have no information on whether there is any substantial spatial variation in the relationships being examined – any such information is lost in the analysis. The situation is akin to being given the information that 'the average temperature in the USA on March 1, 1999 was 15 degrees Celsius'. This is a 'global' statistic in that it provides information about the study area in general but not necessarily about any specific part of it. As such, it is of limited use and disguises the fact that there were substantial temperature variations across the USA on this date. The question spatial analysts need to address is: 'Are there similar spatial variations in analytical results which are being hidden by global statistics?' Until recently, spatial analysts have generally reported globally averaged statistics, such as a single regression parameter estimated from a whole data set, and have ignored any potential spatial non-stationarity in relationships.

   In this chapter, we explore the recent and potentially powerful movement within spatial analysis, termed 'local analysis' or 'local modelling', where the focus of attention is on testing for the presence of *differences* across space rather than on assuming that such differences do not exist. The movement encompasses the dissection of global statistics into their local constituents; the concentration on local exceptions rather than the search for global regularities; and the production of local or mappable statistics rather than on global or 'whole-map' values (Openshaw et al., 1987). As Jones and Hanham (1995) recognize, this trend is important not only because it brings issues of space to the fore in spatial data analysis and spatial modelling, but also because it refutes the rather naive criticism that quantitative geography is unduly concerned with the search for global generalities and 'laws'. Quantitative geographers are increasingly concerned with the development of statistical techniques aimed at the local rather than the global. This shift in emphasis also reflects the increasing availability of large and complex spatial data sets in which spatial variations in relationships are likely to be more apparent.

Understandably, local methods of analysis are important to GIS because they produce values of a statistic for each location and these values can be displayed using the mapping capabilities of GIS (see Chapter 3). It could even be claimed that some of the impetus for the development of local forms of analysis has stemmed from the growing interest in integrating advanced forms of spatial analysis and GIS (Fotheringham and Charlton, 1994; Fotheringham, 1994; Fotheringham and Rogerson, 1993). However, the shifting emphasis from global to local spatial analysis is much broader than that in the GIS literature and involves primarily quantitative geographers and spatial statisticians (*inter alia*, Casetti, 1972; Openshaw et al., 1987; Sampson and Guttorp, 1992; Anselin, 1995; Ord and Getis, 1995). A comparison of the properties of local and global statistics is provided in Table 5.1.

5.2 The nature of local variations in relationships

Consider a frequently encountered aim of data analysis: that of understanding the nature of the distribution of one variable in terms of the distributions of other variables. By far the most popular statistical technique for this purpose is that of regression analysis (Draper and Smith, 1981; Berry and Feldman, 1985; Graybill and Iyer, 1994). In spatial data analysis, the data about which relationships are to be examined are related to spatial units and are used to estimate a single, or 'global', regression equation so that the relationships being examined are assumed to be stationary over space. That is, for each relationship a single parameter estimate is implicitly assumed to depict the nature of that relationship for all points within the entire study area. Clearly, any relationship which is not stationary over space, and which is said to exhibit *spatial non-stationarity*, will not be modelled particularly well by a single parameter estimate and indeed this global estimate may be locally very misleading.

There are at least three reasons to suspect that relationships will vary over space.

Table 5.1 **Characteristics of global and local statistics**

Global	Local
Usually single valued	Multivalued
Assumed invariant over space	Varies over space
Emphasize similarities over space	Emphasize differences across space
Non-mappable	Mappable
('GIS-unfriendly')	('GIS-friendly')
Used to search for regularities	Used to search for exceptions or local 'hotspots'
Aspatial or spatially limited	Spatial

The first and simplest is that there will inevitably be spatial variations in observed relationships caused by random sampling variations. The contribution of this source of spatial non-stationarity is not usually of great interest in itself but it does need to be recognized and accounted for if we are to identify other, more interesting, sources of spatial non-stationarity. That is, we are only interested in relatively large variations in parameter estimates that are unlikely to be due to sampling variation alone.

The second reason is that the relationships might be intrinsically different across space. Perhaps, for example, there are spatial variations in people's attitudes or preferences or there are different administrative, political or other contextual issues that produce different responses to the same stimuli over space. It is difficult to conjecture an example of this cause of spatial non-stationarity in physical geography where the relationships being measured tend to be governed by laws of nature. The idea that human behaviour can vary intrinsically over space is consistent with post-modernist beliefs on the importance of place and locality as frames for understanding such behaviour. Those who hold such a view sometimes criticize quantitative analysis in geography as having little relevance to 'real-world' situations where relationships are very complex and possibly highly contextual. Local statistical indicators address this criticism by recognizing such complexity and attempting to describe it.

The third reason why relationships might exhibit spatial non-stationarity is that the model from which the relationships are measured is a gross misspecification of reality and that one or more relevant variables are either omitted from the model or represented by an incorrect functional form. This view, more in line with the positivist school of thought, assumes a global statement of behaviour can be made (and hence is applicable to relationships in physical as well as human geography) but that the structure of our model is not sufficiently well formed to allow us to make it. In a nutshell, can all contextual effects be removed by a better specification of individual-level effects (Hauser, 1970)? If model misspecification is the cause of parametric instability, the calculation and subsequent mapping of local statistics is useful in order to understand the nature of the misspecification more clearly.

Given the potential importance of local statistics and local models to the understanding of spatial processes, and given the development of what are known as variable parameter models (VPMs) in aspatial contexts (Maddala, 1977; Casetti, 1997), it is surprising that local forms of spatial analysis are not more frequently encountered. As Jones (1991b, p. 8) states, the global modelling approach 'denies geography and history; everywhere and anytime is basically the same! ... [it] is an impoverished representation of reality, and it is amazing that geographers have been so interested in it.' However, there has been a recent flurry of academic work on local spatial analysis reflecting the calls of Fotheringham (1992), Fotheringham and Rogerson (1993) and Openshaw (1993) for greater attention to be given to this topic. These developments can be divided into those that are focused on local statistics for univariate spatial data, which includes the analysis of point patterns, and those that are focused on multivariate data. We now consider both types of application. A brief mention is also made of local forms of spatial interaction models although a fuller discussion of these is left to Chapter 9.

5.3 Measuring local relationships in univariate data

5.3.1 Local point pattern analysis

The analysis of spatial point patterns has long been an important concern in geographical enquiry and a good overview is contained in Boots and Getis (1988); it is also the subject matter of Chapter 6. There has been interest in the topic in several disciplines (*inter alia*, Stone, 1988; Doll, 1989; Gardner, 1989; Besag and Newall, 1991) and particularly in the study of spatial patterns of disease (Marshall, 1991). Until relatively recently, however, most applications of spatial point pattern analysis involved the calculation of some global statistic that described the whole point pattern and from which a conclusion was reached related to the clustered, dispersed or random nature of the whole pattern. Clearly, such analyses are flawed in that any spatial variations in the point pattern are subsumed in the calculation of the average or global statistic. In many instances, particularly in the study of disease, such an approach would appear to be contrary to the purpose of the study, namely to identify any local anomalies.

One of the first developments for the local analysis of point patterns was the geographical analysis machine (GAM) developed by Openshaw et al. (1987) and described in Chapter 6. The technique of the GAM has been criticized by Besag and Newall (1991) and refined by Fotheringham and Zhan (1996) but the basic components remain the same and have great practical application. These are:

1   a method for defining sub-regions of the data;
2   a means of describing the point pattern within each of these sub-regions;
3   a procedure for identifying sub-regions with anomalous point patterns;
4   a device for displaying the sub-regions with anomalous point patterns.

The basic idea suggested by Fotheringham and Zhan (1996) is very simple and serves to demonstrate the interest in the local quite well. Within the study region containing a spatial point pattern, randomly select a location and then randomly select a radius of a circle to be centred at that location. Within this random circle count the number of points and compare this observed value with an expected value based on an assumption about the process generating the point pattern (usually that it is random). Ideally, the population at risk should be used as a basis for generating the expected value, as shown in Fotheringham and Zhan (1996) who use a Poisson probability model with the observed mean and the population at risk within each circle. Once the observed and expected values have been compared, the circle is drawn on a map of the region if it contains an anomalous number of points. The process is repeated many times until a map is produced containing a set of circles centred on parts of the region where interesting clusters of points appear to be located. An example of this type of analysis is given in Openshaw et al. (1987) to examine spatial anomalies in cancer incidence.

The use of automated cluster detection techniques as described above allows for the possibility that different processes are operating in different parts of the region and that these different processes yield different densities of points even when the underlying at-risk population distribution is taken into account. This is quite different from classical approaches such as various neighbour statistics and quadrat analyses that produce 'whole-map' statistics (Dacey, 1960; King, 1961; Greig-Smith, 1964; Robinson et al., 1971; Tinkler, 1971; Boots and Getis, 1988). The 'GAM style' of analysis concentrates on spatial variations and spatial differences in the location of points and hence produces 'local' rather than 'global' information on spatial point patterns. As with all local statistics, GAM-generated statistics are enhanced by the ability to map the results so that variations over space can be easily visualized.

5.3.2 *Other local measures of univariate spatial relationships*

Despite the existence of one or two early papers which emphasized the importance of local or regional variations in relationships (Chorley et al., 1966; Moellering and Tobler, 1972), most of the work in this area is relatively recent. It can be divided into graphical approaches to local data analysis and the more formal development of local univariate statistics.

The bulk of the research undertaken in exploratory graphical analysis (*inter alia*, Haslett et al., 1991; see also Chapter 4) is essentially concerned with identifying local exceptions to general trends in either data or relationships. Hence, techniques such as linked windows and brushing allow data to be examined interactively so that points appearing as outliers in various statistical displays can be located on a map automatically. Usually this type of graphical interrogation takes place with univariate distributions so that histograms form the basis of the graphics although scatterplots can also be linked to a map display and even 3D spin plots can be used (see Figures 4.11, 4.16, 4.20 and 4.21 for examples). No matter which exploratory technique is used, however, the aim of the analysis is generally to identify unusual data points and the focus is on the exceptions rather than the general trend.

More complex graphical techniques for depicting local relationships in univariate data sets include the spatially lagged scatterplot (Cressie, 1984), the variogram cloud plot (Haslett et al., 1991) and the Moran scatterplot (Anselin, 1996). Examples of the latter two are provided in Figures 5.2 and 5.3 respectively for the spatial distribution depicted in Figure 5.1 (which is the data also used in the assessment of statistical inferential tests for spatial autocorrelation in Chapter 8). The data represent unemployment rates distributed across 26 spatial units. The variogram cloud plot in Figure 5.2 depicts the squared difference in the unemployment rates in zones i and j plotted against the distance between zones i and j. The

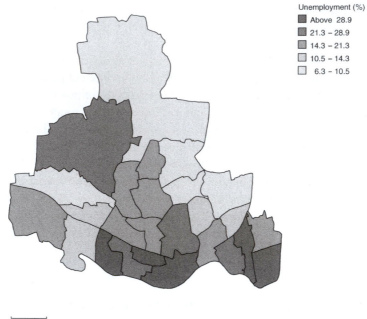

Figure 5.1 **The spatial distribution of unemployment**

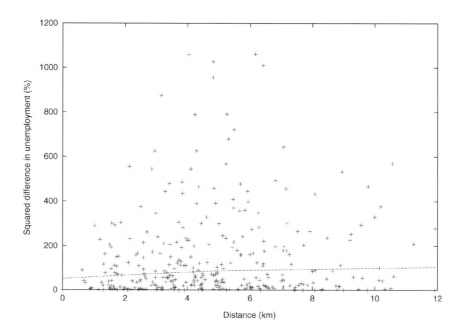

Figure 5.2 **Semi-variogram cloud plot for the unemployment data (curve shown in the Lowess fit)**

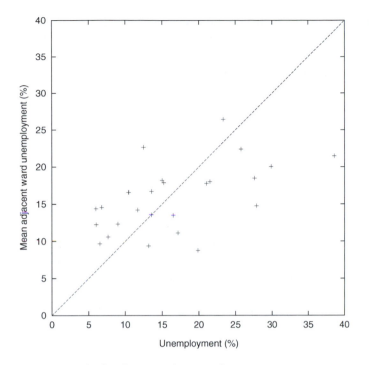

Figure 5.3 **Moran's I-plot for the unemployment data**

plot shows that generally zones closer together have more similar values than those further apart and the strength of this relationship indicates the degree of positive spatial autocorrelation in the data. The development of spatial regression models in Chapter 6 describes one of the consequences of spatial autocorrelation for spatial analysis.

The Moran scatterplot or spatial dependency plot in Figure 5.3 depicts the mean unemployment rate in zones neighbouring zone i plotted against the unemployment rate in zone i. Different definitions of 'neighbouring' can be used in the construction of a Moran scatterplot; here, two zones are classified as neighbours if they share a common boundary. Again, the plot depicts a general trend where neighbouring values are similar, indicating positive spatial autocorrelation. The Moran scatterplot can also be used to depict spatial outliers, defined as zones having very different values of an attribute from their neighbours.

More formally, local versions of global univariate statistics have recently been developed by Getis and Ord (1992), Ord and Getis (1995) and by Anselin (1995; 1998). Getis and Ord (1992) develop a global measure of spatial association inherent within a data set that measures the way in which values of an attribute are clustered in space. A local variant of this global statistic is then formulated to depict trends in the data around each spatial unit. In fact there are two variants of this localized statistic depending on whether or not the unit i around which the clustering is measured is included in the calculation. Unfortunately there is no theory to guide the use of which statistic to use in any particular situation although the difference between the two will typically be very small in most situations where there are large numbers of spatial units.

For the situation where i is *not* included in the calculation

$$G_i = \sum_j w_{ij} x_j \Big/ \sum_j x_j \qquad j \neq i \qquad (5.1)$$

where G_i is the measure of local clustering of attribute x around i, x_j is the value of x at j, and w_{ij} represents the strength of the spatial relationship between units i and j which can be measured as either a binary contiguity variable or a continuous distance-decay measure (the latter generally being preferred although the choice of spatial weighting function is generally subjective). If high values of x tend to be clustered around i, G_i will be high; if low values of x tend to cluster around i then G_i will be low; and no distinct clustering of high or low values of x around i will produce intermediate values of G_i. The expected value of G_i is

$$E(G_i) = w_{i.}/(n-1) \qquad (5.2)$$

where

$$w_{i.} = \sum_j w_{ij} \qquad j \neq i \qquad (5.3)$$

and the variance is

$$\mathrm{var}(G_i) = w_{i.}(n-1-w_{i.})s_i^2/(n-1)^2(n-2)\bar{x}_i^2 \qquad (5.4)$$

where $\bar{x}_i$ represents the mean of the x values excluding the value at i and s_i^2 is the sample estimate of the variance of x, again excluding the value at i. A standard variate can then be defined as

$$Z(G_i) = [G_i - E(G_i)]/[\mathrm{var}(G_i)]^{1/2} \qquad (5.5)$$

For the situation when i is included in the calculation, then the above formulae simplify to

$$G_i^* = \sum_j w_{ij}x_j \bigg/ \sum_j x_j \qquad \text{for all } j \qquad (5.6)$$

where w_{ii} must not equal zero,

$$E(G_i^*) = w_{i.}^* / n \qquad (5.7)$$

where

$$w_{i.}^* = \sum_j w_{ij} \qquad \text{for all } j \qquad (5.8)$$

and

$$\text{var}(G_i^*) = w_{i.}^*(n - w_{i.}^*)s^2 / (n)^2(n-1)\bar{x}^2 \qquad (5.9)$$

and where the mean and sample estimate of the variance are taken over the entire set of x values and so no longer vary with i.

The G_i and G_i^* statistics have been implemented in a GIS environment by Ding and Fotheringham (1992) which has the advantage of a readily available mapping system for the local statistics. The local spatial association statistic allows that different trends in the distribution of one variable might exist over space. In some parts of the study area, for example, high values might be clustered; in other parts there might be a mix of high and low values. Such differences would not be apparent in the calculation of a single global statistic. In their empirical example, Getis and Ord (1992) find several statistically significant local clusters of sudden infant death syndrome in North Carolina although the global statistic fails to identify any significant clustering.

In a similar manner to that of Getis and Ord (1992), Anselin (1995) has recently developed a local variant of Moran's I, a measure of spatial autocorrelation (see Chapter 8 for more detail). Spatial autocorrelation is traditionally measured globally so that the statistic describes an average trend in the way a variable is distributed over space. Where spatial data are distributed so that high values are generally located near to other high values and low values are generally located near to other low values, the data are said to exhibit positive spatial autocorrelation. Where the data are distributed such that high and low values are generally located near each other, the data are said to exhibit negative spatial autocorrelation. Clearly these descriptions are global ones and may not adequately describe the relationships in all parts of the study area. Anselin's development of a localized version of spatial autocorrelation allows spatial variations in the arrangement of a variable to be examined. The localized version of Moran's I is

$$I_i = \frac{(x_i - x^*)\sum_j w_{ij}(x_j - x^*)}{\sum_i (x_i - x^*)^2 / n} \qquad (5.10)$$

where x_i is the observed value of x at location i, x^* is the mean of x, n is the number of observations and w_{ij} represents the strength of the linkage between i and j, usually measured by spatial proximity. The expected value and variance of I_i are given by

$$E(I_i) = -w_{i.}/(n-1) \qquad (5.11)$$

and

$$\text{var}(I_i) = w_{i.}^2 V \qquad (5.12)$$

respectively, where V is the calculation of the variance of I under randomization with w_{ij} replaced throughout by w_i (the formula for the variance of I is given in Chapter 8 in Equations (8.17) to (8.20) and is not repeated here). Again, the local Moran's I statistics can be mapped to display any local variations in spatial autocorrelation. Anselin (1995) presents an application of the localized Moran's I statistic to the spatial distribution of conflict in Africa and Sokal et al. (1998) demonstrate its use on a set of simulated data sets. Other studies of local Moran's I statistic include those of Bao and Henry (1996), Tiefelsdorf and Boots (1997), Tiefelsdorf (1998) and Tiefelsdorf et al. (1998).

5.4 Measuring local relationships in multivariate data

The increasing availability of large and complex spatial data sets has led to a greater awareness that the univariate statistical methods described in Section 5.3 have limited application. There is a need to understand local variations in more complex relationships (see e.g. the attempts of Ver Hoef and Cressie (1993) and Majure and Cressie (1997) to extend the local visual techniques for autocorrelation described above to the multivariate case). In response to this, several attempts have been made to produce localized versions of traditionally global multivariate techniques. For instance, it was recognized quite early in the spatial interaction modelling literature that localized distance-decay parameters would yield more useful information on spatially varying rates of decay than simply estimating a global interaction model (see Fotheringham (1981) for a review and also Chapter

9). From an accumulation of empirical examples of origin-specific parameter estimates, it has proven possible to map trends in parameter estimates that have led to the identification of a severe misspecification bias in the general spatial inter-action modelling formula (Fotheringham, 1984; 1986). It is worth stressing that such misspecification only came to light through an investigation of spatial vari-ations in localized parameters that would be completely missed in the calibration of a global model. This topic forms part of the subject matter of Chapter 9 and more detailed discussion is left to this point.

Perhaps the most pressing challenge, given its widespread use, has been to produce local versions of regression analysis. Examples include spatial adaptive filtering (Foster and Gorr, 1986; Gorr and Olligschlaeger, 1994) which incorporates spatial relationships in an *ad hoc* manner and produces parameter estimates which are difficult to test statistically; random coefficients modelling (Aitkin, 1996); and multilevel modelling (Goldstein, 1987; Westert and Verhoeff, 1997). In the latter two techniques, it is possible to obtain local parameter estimates through Bayes' theorem. However, the application of both frameworks to the analysis of *spatial* data is questionable because no assumption is made about any spatial dependency in the relationships being examined which would seem unrealistic of most spatial processes. Brunsdon et al. (1999b) provide an example of the application of random coefficient models to spatial data and Jones (1991a; 1991b) provides examples of multilevel modelling with spatial data. The latter relies heavily on a predefined set of hierarchically arranged spatial units in which the continuous nature of spatial processes, both within and between levels, is essentially ignored. The only notion of spatial arrangement is one of 'containment', where the coefficients for smaller areal units are 'nested' inside those for larger ones.

Two further examples of local regression techniques which would seem to have greater potential application to spatial data are a spatial variant of the expansion method (Casetti, 1972; Brown and Jones, 1985; Jones and Casetti, 1992) and geographically weighted regression (Brunsdon et al., 1996; 1998a; 1998b; Fother-ingham et al., 1996; 1997a; 1997b; 1998). Following a brief discussion of multi-level modelling, these latter two techniques are described and applications of both to spatial data analysis are then presented.

5.4.1 Multilevel modelling

The typical application of multilevel modelling attempts to separate the effects of personal characteristics and place characteristics (contextual effects) on behaviour, a problem which arises in many different situations (inter alia, Hauser, 1970; Meyer and Eagle, 1982; Brown and Kodras, 1987; Brown and Goetz, 1987). Modelling spatial behaviour purely at the individual level is prone to the atomistic fallacy, missing the context in which individual behaviour occurs (Alker, 1969), whereas modelling behaviour at the aggregate level is prone to the ecological fallacy that the results might not apply to individual behaviour (Robinson, 1950). Multilevel modelling tries to avoid both these problems by combining an individual-level model representing disaggregate behaviour with a macro-level model representing contextual variations in behaviour.

To see how multilevel modelling accomplishes this, consider modelling the behaviour of individual i in place j. At the individual level,

$$y_i = \alpha + \beta x_i + \varepsilon_i \qquad (5.13)$$

where y_i represents some facet of the behaviour of individual i; x_i is an attribute of individual i that affects this behaviour; α and β are parameters to be estimated and ε is a normally distributed random term representing variation in response around a mean estimated from $\alpha + \beta x_i$. Alternatively, at the place or aggregate level,

$$y_j = \gamma + \delta x_j + \epsilon_j \qquad (5.14)$$

where y_j represents some facet of the behaviour of a group of individuals at place j; x_j is an attribute of place j; γ and δ are parameters to be estimated; and ϵ_j represents the level of the random term at the place level.

In multilevel modelling, these two equations are combined to produce a model of the form

$$y_{ij} = \alpha_j + \beta_j x_{ij} + e_{ij} \qquad (5.15)$$

in which y_{ij} represents the behaviour of individual i living in place j; α_j and β_j are place-specific parameters where

$$\alpha_j = \alpha + \mu_{j\alpha} \qquad (5.16)$$

and

$$\beta_j = \beta + \mu_{j\beta}; \qquad (5.17)$$

and e_{ij} represents the random error associated with individual i in place j.

Each place-specific parameter is therefore viewed as consisting of an average value plus a random component. It is clear from these expressions that multilevel modelling is a more sophisticated version of the use of dummy variables in regression to obtain place-specific parameters (see Johnston et al., 1988, p. 179, for an example of this technique used to obtain region-specific parameters). Substituting (5.16) and (5.17) into (5.15) yields the multilevel model

$$y_{ij} = \alpha + \beta x_{ij} + (e_{ij} + \mu_{j\alpha} + \mu_{j\beta}x_{ij}) \qquad (5.18)$$

which, because it contains three random components, cannot be calibrated by OLS regression unless $\mu_{j\alpha}$ and $\mu_{j\beta}$ are zero, and specialized software is needed such as Mln (Rasbash and Woodhouse, 1995). Place-specific parameter estimates can be obtained by estimating separate variance effects and substituting into Equations (5.16) and (5.17).

Refinements to the basic multilevel model described above can be made. For instance, place attributes can be included in the specifications for α_j and β_j so that, for example, if one place attribute, z_j, affects α_j and β_j then

$$\alpha_j = \alpha + \phi z_j + \mu_{j\alpha} \qquad (5.19)$$

and

$$\beta_j = \beta + \lambda z_j + \mu_{j\beta} \qquad (5.20)$$

Substituting (5.19) and (5.20) into (5.15) yields

$$y_{ij} = \alpha + \beta x_{ij} + \lambda z_j x_{ij} + \phi z_j + (e_{ij} + \mu_{j\alpha} + \mu_{j\beta}x_{ij}) \qquad (5.21)$$

Other refinements include extending the number of levels in the hierarchy beyond two so that threefold and n-fold hierarchies can be modelled (Jones et al., 1996) and the development of cross-classified multilevel models where each lower unit can nest into more than one higher-order unit (Goldstein, 1994).

There are several examples of multilevel modelling applied to spatial data and overviews are given by Jones (1991a; 1991b). Applications include those of Jones (1997) on voting behaviour; Verheij (1997) and Duncan et al. (1996) on geographical variations in the utilization of health services and health-related behaviour; Jones and Bullen (1993) on spatial variations in house prices; Smit (1997) on commuting patterns; and Duncan (1997) on spatial patterns of smoking behaviour. However, each of these studies relies on an a priori definition of a discrete set of spatial units at each level of the hierarchy. While this may not be a problem in many aspatial applications, such as the definition of what constitutes the sets of public and private transportation options, or what constitutes the sets of brands of decaffeinated and regular coffees, it can pose a problem in many spatial contexts. The definition of discrete spatial entities in which spatial behaviour is modified by attributes of those entities obviously depends on such entities being identified by the modeller and this is not always possible with spatial processes. It also implies that the nature of whatever spatial process is being modelled is discontinuous. That is, it is assumed that the process is modified in exactly the same way throughout a particular spatial unit but it is modified in a different way as soon as the boundary of that spatial unit is crossed. Most spatial processes do not operate in this way because the effects of space are continuous. Hence, imposing a discrete set of boundaries on most spatial processes is unrealistic.[1] There are exceptions to this, however, such as where administrative boundaries enclose regions in which a policy that affects the behaviour of individuals is applied evenly throughout the region and where such policies vary from region to region. Examples of spatial units where this type of behaviour modification might occur are the states in the USA, education and health districts in the UK, and the individual countries of the European Union.

5.4.2 The spatial expansion method

The expansion method (Casetti, 1972; 1997; Jones and Casetti, 1992) attempts to measure parameter 'drift'. In this framework, parameters of a global model can be made functions of other attributes, including location, so that *trends* in parameter estimates over space can be measured (Brown and Jones, 1985; Brown and Kodras, 1987; Brown and Goetz, 1987; Fotheringham and Pitts, 1995; Eldridge and Jones, 1991). Initially, a global model is proposed such as

$$y_i = \alpha + \beta x_{i1} + \cdots + \tau x_{im} + \varepsilon_i \tag{5.22}$$

where y represents a dependent variable; the x are independent variables, of which there are m; α, β, ..., τ represent parameters to be estimated; ε represents an error term; and i represents a point in space at which observations on the y and x are recorded. This global model can be expanded by allowing each of the parameters to be functions of other variables. Whilst most applications of the expansion method (see Jones and Casetti, 1992) have undertaken aspatial expansions, it is relatively straightforward to allow the parameters to vary over geographical space so that, for example,

$$\alpha_i = \alpha_0 + \alpha_1 u_i + \alpha_2 v_i \tag{5.23}$$

$$\beta_i = \beta_0 + \beta_1 u_i + \beta_2 v_i \tag{5.24}$$

and

$$\tau_i = \tau_0 + \tau_1 u_i + \tau_2 v_i \tag{5.25}$$

where u_i and v_i represent the spatial coordinates of location i. Equations (5.23)–(5.25) represent very simple linear expansions of the global parameters over space, but more complex, non-linear, expansions such as

$$\alpha_i = a_0 + a_1 u_i + a_2 v_i + a_3 u_i^2 + a_4 v_i^2 + a_5 u_i v_i \qquad (5.26)$$

can easily be accommodated although the interpretation of individual parameter estimates is then usually less obvious.

Once a suitable form for the expansion has been chosen, the original parameters in the basic model are replaced with their expansions. For instance, if it is assumed that parameter variation over space can be captured by the simple linear expansions in Equations (5.23)–(5.25), the expanded model would be

$$y_i = a_0 + a_1 u_i + a_2 v_i + \beta_0 x_{i1} + \beta_1 u_i x_{i1} + \beta_2 v_i x_{i1}$$

$$+ \cdots + \tau_0 x_m + \tau_1 u_i x_{im} + \tau_2 v_i x_{im} + \varepsilon_i \qquad (5.27)$$

Equation (5.27) can then be calibrated by OLS regression (or maximum likelihood estimation if the model is a Poisson or logit regression or if it contains spatial autoregressive terms) to produce estimates of the parameters which are then fed back into Equations (5.23)–(5.25) to obtain spatially varying parameter estimates. These estimates, being specific to location i, can then be mapped to display spatial variations in the relationships represented by the parameters.

The expansion method, despite its importance in highlighting the concept that relationships might vary over space and that, consequently, the parameters of regression models applied to spatial data might exhibit spatial non-stationarity, has several limitations. One is that the technique is restricted to displaying *trends* in relationships over space with the complexity of the measured trends being dependent upon the complexity of the expansion equations. Clearly the spatially varying parameter estimates obtained through the expansion method might obscure important local variations to the broad trends represented by the expansion equations. A second is that the form of the expansion equations needs to be assumed a priori although more flexible functional forms than those shown above could be used. A third is that the expansion equations must be deterministic to remove problems of estimation in the terminal model. All three problems are overcome in the second method of producing local estimates of regression parameters, namely geographically weighted regression.

5.4.3 Geographically weighted regression

Consider a global regression model given by

$$y_i = a_0 + \sum_k a_k x_{ik} + \varepsilon_i \qquad (5.28)$$

In the calibration of this model, one parameter is estimated for the relationship between each independent variable and the dependent variable and this relationship is assumed to be constant across the study region. The estimator for (5.28) is

$$\mathbf{a} = (\mathbf{X}^T\mathbf{X})^{-1}\mathbf{X}^T\mathbf{y} \tag{5.29}$$

where **a** represents the vector of global parameters to be estimated, **X** is a matrix of independent variables with the elements of the first column set to 1, and **y** represents a vector of observations on the dependent variable. Geographically weighted regression (GWR) is a relatively simple technique that extends the traditional regression framework of Equation (5.28) by allowing local rather than global parameters to be estimated so that the model is rewritten as

$$y_i = a_0(u_i, v_i) + \sum_k a_k(u_i, v_i)x_{ik} + \varepsilon_i \tag{5.30}$$

where (u_i, v_i) denotes the coordinates of the ith point in space and $a_k(u_i, v_i)$ is a realization of the continuous function $a_k(u, v)$ at point i (Brunsdon et al., 1996; 1998a; 1998b; Fotheringham et al., 1996; 1997a; 1997b; 1998; 1999). That is, we allow there to be a continuous surface of parameter values and measurements of this surface are taken at certain points to denote the spatial variability of the surface. Note that the global model in Equation (5.28) is a special case of the GWR model represented by Equation (5.30) in which the parameter surface is assumed to be constant over space. Thus the GWR Equation (5.30) recognizes that spatial variations in relationships might exist and provides a way in which they can be measured.

In the calibration of the GWR model it is assumed that observed data near to point i have more of an influence in the estimation of the $a_k(u_i, v_i)$ than do data located farther from i. In essence, the equation measures the relationships inherent in the model *around each point i*. Hence weighted least squares estimation provides a basis for understanding how GWR operates. In GWR an observation is weighted in accordance with its proximity to point i so that the weighting of an observation is no longer constant in the calibration but varies with i. Data from observations close to i are weighted more than data from observations farther away. This is shown in Figure 5.4 where a spatial kernel is placed over each calibration point and the data around that point are weighted according to the distance-decay curve displayed by the kernel (examples of spatial kernels are shown in Chapter 2).

Algebraically, the GWR estimator is

$$\mathbf{a}(u_i, v_i) = (\mathbf{X}^T\mathbf{W}(u_i, v_i)\mathbf{X})^{-1}\mathbf{X}^T\mathbf{W}(u_i, v_i)\mathbf{y} \tag{5.31}$$

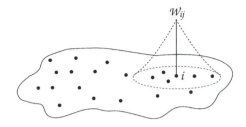

Figure 5.4   **Example of a spatial kernel in GWR**

where $\mathbf{W}(u_i, v_i)$ is an n by n matrix whose off-diagonal elements are zero and whose diagonal elements denote the geographical weighting of observed data *for point i*. That is,

$$\mathbf{W}(u_i, v_i) = \begin{pmatrix} w_{i1} & 0 & 0 & \cdots & 0 \\ 0 & w_{i2} & 0 & \cdots & 0 \\ 0 & 0 & w_{i3} & \cdots & 0 \\ \vdots & \vdots & \vdots & \cdots & \vdots \\ 0 & 0 & 0 & \cdots & w_{in} \end{pmatrix} \qquad (5.32)$$

where w_{in} denotes the weight of the data at point n on the calibration of the model around point i. Clearly, these weights will vary with i which distinguishes GWR from traditional weighted least squares where the weighting matrix is constant. Below we describe how these weights can be defined.[2]

There are parallels between GWR, kernel regression and drift analysis of regression parameters (DARP) (Cleveland, 1979; Cleveland and Devlin, 1988; Casetti, 1982). In kernel regression and DARP, $\mathbf{y}$ is modelled as a non-linear function of $\mathbf{X}$ by weighting data in attribute space rather than geographical space. That is, data points more similar to x_i are weighted more heavily than data points that are less similar and the output is a set of localized parameter estimates in x space. However, Casetti and Jones (1983) do provide a limited spatial application of DARP that is very similar in intent to GWR although it lacks a formal calibration mechanism and significance testing framework and so is treated by the authors as a rather limited heuristic method.

It should be noted that as well as producing localized parameter estimates, the GWR technique described above will produce localized versions of all standard regression diagnostics including goodness-of-fit measures such as r^2. The latter can be particularly informative in understanding the application of the model being calibrated and for exploring the possibility of adding further explanatory variables to the model. It is also useful to note that the points for which parameters are locally estimated in GWR need not be the points at which data are collected: estimates of parameters can be obtained for any location. Hence, in systems with very large numbers of data points, GWR estimation of local parameters can take place at predefined intervals such as at the intersections of a grid placed over the study region. Not only does this reduce computing time but it can also be beneficial for mapping the results.

Until this point, it has merely been stated in GWR that $\mathbf{W}(u_i, v_i)$ is a weighting scheme based on the proximity of i to the sampling locations around i without an explicit relationship being stated. The choice of such a relationship is now considered. First, consider the implicit weighting scheme of the OLS framework in Equation (5.28). In this case

$$\mathbf{W}(u_i, v_i) = \begin{pmatrix} 1 & 0 & 0 & \cdots & 0 \\ 0 & 1 & 0 & \cdots & 0 \\ 0 & 0 & 1 & \cdots & 0 \\ \vdots & \vdots & \vdots & \cdots & \vdots \\ 0 & 0 & 0 & \cdots & 1 \end{pmatrix} \tag{5.33}$$

That is, the global model is equivalent to a local model in which each observation has a weight of unity so that there is no spatial variation in the estimated parameters. An initial step towards weighting based on locality might be to exclude from the calibration of the model at location i observations that are further than some distance d from i. Suppose i represents a calibration point and j represents a data point. Then, the weights could be defined as

$$\begin{aligned} w_{ij} &= 1 && \text{if} \quad d_{ij} \leqslant d \\ w_{ij} &= 0 && \text{otherwise} \end{aligned} \tag{5.34}$$

so that the diagonal elements of (5.32) would be 0 or 1 depending on the whether or not the criterion in (5.34) is met. The use of (5.34) to define the diagonal elements of (5.32) would simplify the calibration procedure since for every point for which coefficients are to be computed only a subset of the sample points needs to be included in the regression model. Examples of the use of a discrete weighting function in GWR are provided in Fotheringham et al. (1996) and Charlton et al. (1996).

However, the spatial weighting function in (5.34) does not reflect actual geographical processes very well because it suffers from the problem of discontinuity. As i varies around the study area, the regression coefficients could change drastically as one sample point moves into or out of the circular buffer around i which defines the data to be included in the calibration for location i. Although sudden changes in the parameters over space might genuinely occur, in this case changes in their estimates would be artefacts of the arrangement of sample points, rather than any underlying process in the phenomena under investigation. One way to combat this is to specify w_{ij} as a continuous function of d_{ij}, the distance between i and j. One obvious choice is to define the diagonal elements of (5.32) by

$$w_{ij} = \exp(-d_{ij}^2/h^2) \tag{5.35}$$

where h is referred to as the bandwidth. If i and j coincide (i.e. i also happens to be a point in space at which data are observed), the weighting of data at that point will be unity. The weighting of other data will decrease according to a Gaussian curve as the distance between i and j increases. In the latter case the inclusion of data in the calibration procedure becomes 'fractional'. For example, in the calibration of a model for point i, if $w_{ij} = 0.5$ then data at point j contribute only half the weight in the calibration procedure as data at point i itself. For data a long way from i the weighting will fall virtually to zero, effectively excluding these observations from the estimation of parameters for location i.

To this point, it is assumed that the spatial weighting function in (5.35) is applied equally at each calibration point. In effect, this is a global statement of the weight–distance relationship and as such it suffers from the potential problem that in some parts of the region, where data are sparse, the local regressions might be based on relatively few data points. To offset this potential problem, spatially adaptive weighting functions can be incorporated into GWR. These would have relatively small bandwidths in areas where the data points are densely distributed and relatively large bandwidths where the data points are sparsely distributed. An example of how such a spatially adaptive kernel would operate is shown in Figure 5.5.

The following three weighting functions which each define the diagonal elements of (5.32) produce spatially adaptive kernels. The first is based on nearest neighbours:

$$w_{ij} = \begin{cases} [1 - (d_{ij}/h_i)^2]^2 & \text{if } d_{ij} < h_i \\ 0 & \text{otherwise} \end{cases} \tag{5.36}$$

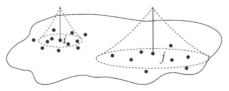

Point i is in a relatively dense cluster of data and the kernel is steep

Point j is in a relatively sparse cluster of data and the kernel is gentle

Figure 5.5 **Example of an adaptive spatial kernel in GWR**

where h_i is the Nth nearest-neighbour distance from i. The second is based on ranked distances:

$$w_{ij} = \exp(-\delta R_{ij}) \tag{5.37}$$

where R_{ij} is the rank of the distance data point j is from calibration point i. The third constrains the sum of the weights for any calibration point to be a constant:

$$\sum_j w_{ij} = k \qquad \text{for all } i \tag{5.38}$$

All three functions will automatically ensure that the weighting kernels will expand in areas where the data are scarce and contract in areas that are data rich.

Whatever the specific weighting function employed, the essential idea of GWR is that for each point i sampled observations near to i have more influence in the estimation of the parameters for point i than do sampled observations farther away. Obviously, whichever weighting function is selected, the estimated parameter surfaces will be, in part, functions of the definition of that weighting function. In (5.34), for example, the larger is d, the more data points will have full weight in the local regression and the closer will be the local model solution to that of the global one. Obviously, if a value of d is chosen such that no value of d_{ij} exceeds this value, then all data points will have full weight in all the local regressions and the local and global models will produce the same results. Equivalently, in (5.35) as h tends to infinity (no distance decay), the weights tend to one for all pairs of points so that the estimated parameters become uniform and GWR becomes equivalent to OLS. Conversely, as the bandwidth becomes smaller, the parameter estimates will increasingly depend on observations in close proximity to i and hence will have increased variance. The problem is therefore how to select an appropriate kernel and an appropriate bandwidth for that kernel. Actually, the choice of kernel is relatively unimportant as long as it is a continuous function in which the weights decrease as distance increases; the selection of an appropriate bandwidth is much more important (Brunsdon et al., 1996; Fotheringham et al., 1997b; 1999).

To demonstrate a technique for the estimation of the bandwidth, consider the selection of h in (5.35). One possibility is to choose h on a 'least-squares' criterion; that is, to minimize the quantity z

$$z = \sum_{i=1,n} [y_i - y_i(h)]^2 \tag{5.39}$$

where $y_i(h)$ is the fitted value of y_i using a bandwidth of h. In order to find the fitted value of y_i it is necessary to estimate the $a_k(u_i, v_i)$ at each of the sample points and then combine these with the x values at these points. However, when minimizing the sum of squared errors suggested above, a problem is encountered. Suppose h is made very small so that the weighting of all points except for i itself becomes negligible. Then the fitted values at the sampled points will tend to the *actual* values, so that the value of (5.39) becomes zero. This suggests that under such an optimizing criterion the value of h tends to zero, but clearly this degenerate case is not helpful. First, the parameters of such a model are not defined in this limiting case and, secondly, the estimates will fluctuate wildly throughout space in order to give locally good fitted values at each location.

A solution to this problem is a *cross-validation* (CV) approach suggested for local regression by Cleveland (1979) and for kernel density estimation by Bowman (1984). Here, a score of the form

$$z = \sum_{i=1,n}[y_i - y_{\neq i}(h)]^2 \qquad (5.40)$$

is used where $y_{\neq i}(h)$ is the fitted value of y_i with the observations for point i omitted from the calibration process. This approach has the desirable property of countering the 'wrap-around' effect since, when h becomes very small, the model is calibrated only on samples near to i and not at i itself. Plotting the CV score against the required parameter of whatever weighting function is selected will therefore provide guidance on selecting an appropriate value of that parameter. The process can be automated by maximizing the CV score using an optimization technique such as a golden section search, assuming the cross-validation function is reasonably well behaved (Greig, 1980).

Once a weighting function has been selected and calibrated, the output from GWR will be a set of local parameter estimates for each relationship in the model. Because these local estimates are all associated with specific locations, each set can be mapped to show the spatial variation in the measured relationship. Similarly, local measures of standard errors and goodness-of-fit statistics are obtained. Given that, as we identified earlier, there might be different causes of spatial non-stationarity, one of which is random sampling variation, it is useful to ask the question: 'Does the set of local parameter estimates exhibit significant spatial variation?' Here we describe an experimental procedure to answer this question although the reader is also referred to the theoretical tests described by Brunsdon et al. (1999a).[3]

The variability of the local estimates can be used to examine the plausibility of the stationarity assumption held in traditional regression. In general terms, this could be thought of as a variance measure. For a given k, suppose $a_k(u_i, v_i)$ is the GWR estimate of $a_k(u, v)$. Suppose we take n values of this parameter estimate (one for each point i within the region); then an estimate of variability in the parameter is given by the standard deviation of the n parameter estimates. This statistic will be referred to as s_k. The next stage is to determine the sampling distribution of s_k under the null hypothesis that the global model in Equation (5.28) holds. Under the null hypothesis, any permutation of (u_i, v_i) pairs amongst the geographical sampling points i are equally likely to occur. Thus, the observed value of s_k could be compared with the values obtained from randomly rearranging the data in space and repeating the GWR procedure. The comparison between the observed s_k value and those obtained from a large number (99 in this case) of randomization distributions forms the basis of the significance test. Making use of the Monte Carlo approach, it is also the case that selecting a subset of random permutations of (u_i, v_i) pairs amongst the i and computing s_k will also give a significance test when compared with the observed statistics.

More details on GWR can be found in Brunsdon et al. (1996; 1998a; 1998b; 1999a) and in Fotheringham et al. (1996; 1997a; 1997b; 1998). More specifically, further information on theoretical significance tests for GWR models can be found in Brunsdon et al. (1999a); on calibration issues and variance–bias trade-off in GWR parameter estimates in Fotheringham et al. (1998); the use of GWR as an alternative method of producing local spatial autocorrelation measures in Brunsdon et al. (1998a); and on GWR and scale issues in Fotheringham et al. (1999). Copies of the programs and sample data sets to run GWR can be obtained at the following web site: *http://www.ncl.ac.uk/geography/GWR*

5.5 An empirical comparison of the spatial expansion method and GWR

5.5.1 The data

An example of the application of both the spatial expansion method and GWR is now described using data on the 1991 spatial distribution of limiting long-term illness (LLTI) which is a self-reported variable asked in the UK Census of Population. It encompasses a variety of severe illnesses such as respiratory diseases, multiple sclerosis, heart disease, severe arthritis as well as physical disabilities that prevent people from being in the labour market. The study area encompasses 605 spatial units (census wards) in four administrative counties in North-East England: Tyne and Wear, Durham, Cleveland and North Yorkshire. Tyne and Wear is a heavily populated service and industrial conurbation in the northern part of the study area and is centred on the city of Newcastle. To the south, Durham has been heavily dependent on coal mining in the eastern half of the county with the western half being predominantly rural. Cleveland, to the south-east, is a largely urban, industrial area with heavy petrochemical and engineering works clustered around the Tees estuary and centred on Middlesbrough. North Yorkshire, to the south, is a predominantly rural and fairly wealthy county with few urban areas. The distribution of urban areas throughout the region is shown in Figure 5.6.

The spatial distribution of a standardized measure of LLTI (defined as the percentage of individuals aged 45–65 living in a household where LLTI is reported) throughout the study region is shown in Figure 5.7. As one might expect,

Figure 5.6 **Urban areas in the study region**

the LLTI variable tends to be higher in the industrial regions of Tyne and Wear, east Durham and Cleveland and lower in the rural areas of west Durham and North Yorkshire.

5.5.2 Global regression model results

To model the spatial distribution of LLTI in this region, the following global regression model was constructed:

$$\text{LLTI}_i = a_0 + a_1\text{UNEM}_i + a_2\text{CROW}_i + a_3\text{SPF}_i + a_4\text{SC1}_i + a_5\text{DENS}_i \quad (5.41)$$

where LLTI is the age-standardized measure of LLTI described above; UNEM is the proportion of economically active males and females who are unemployed (the denominator in this variable does not include those with LLTI who are not classed as being economically active); CROW is the proportion of households

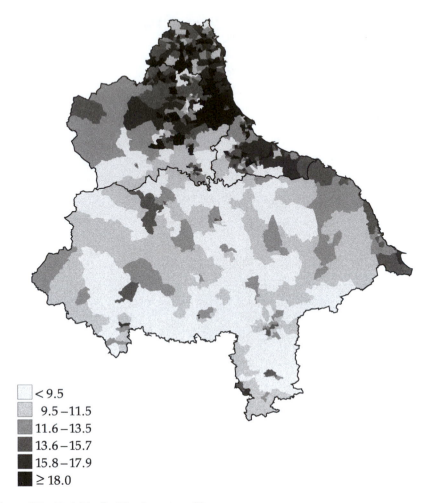

	< 9.5
	9.5 – 11.5
	11.6 – 13.5
	13.6 – 15.7
	15.8 – 17.9
	≥ 18.0

Figure 5.7   **Variable: limiting long-term illness**

whose inhabitants are living at a density of over one person per room; SPF is the proportion of households with single parents and children under 5; SC1 is the proportion of residents living in households with the head of household in social class 1 (employed in professional, non-managerial occupations); and DENS is the density of population in millions per square kilometre. This last variable discriminates particularly well between urban and rural areas. The model is guided by the findings of Rees (1995) in his examination of LLTI at a much coarser spatial resolution (English and Welsh counties and Scottish regions). The data are extracted from the 1991 UK Census of Population Local Base Statistics. The areal units used are census wards that contain on average approximately 200 households per ward. Using these data, the calibrated form of the global model is

$$\text{LLTI}_i = 3.8 + 92.6\text{UNEM}_i + 31.1\text{CROW}_i - 3.5\text{SPF}_i - 22.5\text{SC1}_i - 5.6\text{DENS}_i$$
$$(1.3) \quad (3.4) \qquad\quad (3.9) \qquad\quad (2.3) \qquad (4.1) \qquad (2.5)$$

$$(5.42)$$

where the numbers in brackets represent t-statistics and the r^2 value associated with the regression is 0.76. The results suggest that across the study region LLTI is positively related to unemployment levels and crowding. The former relationship reflects perhaps that the incidence of LLTI is related to both social and employment conditions in that the areas of higher unemployment tend to be the wards with lower incomes and higher proportions of declining heavy industries. The latter relationship suggests a link between LLTI and social conditions with levels of LLTI being higher in areas with high levels of overcrowding. LLTI is negatively related to the proportion of professionally employed people in a ward and to population density. The negative relationship between LLTI and SC1 supports the hypothesis that LLTI is more prevalent in poorer areas that have fewer people in professional occupations. It also reflects the fact that industrial hazards, which are a factor in the incidence of LLTI, are less likely to be faced by people in professional occupations. The negative relationship between LLTI and DENS is somewhat counter-intuitive in that it suggests that LLTI is greater in less densely populated areas, *ceteris paribus*. The nature of this latter relationship is explored in greater detail below in the discussion of the GWR results. Only the single-parent family variable is not significant (at 95%) in the global model.

To this point, the empirical results and their interpretations are typical of those found in standard regression applications; parameter estimates are obtained which are assumed to describe relationships that are invariant across the study region. We now describe how the spatial expansion method and GWR can be used to examine the validity of this assumption and explore in greater detail spatial variations in the relationships described above.

5.5.3 Spatial expansion method results

Each of the six parameters in the global model given in Equation (5.41) was expanded in terms of linear functions of space as depicted in Equations (5.23)–(5.25) (the results of more complicated quadratic expansions are given in Fotheringham et al., 1998). The results are shown in Table 5.2 along with the parameter estimates from the global model.

The results suggest that significant spatial variation in three parameters exists: the unemployment parameter, the social class parameter and the density parameter all appear to decrease significantly moving from west to east across the region (the unemployment becomes less positive and the social class and density parameters become more negative). The other parameters apparently exhibit no significant spatial variation. To see this more clearly, values of the x, y coordinates for each spatial unit (the 605 census wards) are input into the calibrated expansion equations

Table 5.2 **Linear expansion method results**

Variable	Global	Linear EM
Constant	3.8*	5.2
Constant. u_i		0.000064
Constant. v_i		−0.000067
Unem	93*	430*
Unem. u_i		−0.00048*
Unem. v_i		−0.00023
Crow	31*	−220
Crow. u_i		0.00017
Crow. v_i		0.00034
SC1	−23*	210
SC1. u_i		−0.00028*
SC1. v_i		−0.0002
Dens	−5.6*	140
Dens. u_i		−0.00016*
Dens. v_i		−0.00013
SPF	−3.5	55
SPF. u_i		−0.000038
SPF. v_i		−0.000083
r^2	0.76	0.80
DoF	589	577

*Indicates that the parameter estimate is significantly different from zero at the 95% level.

for each parameter and the resulting locally varying parameter estimates are mapped. These spatial distributions are shown below for the three parameters discussed above plus the intercept term.

The spatially varying intercept is shown in Figure 5.8. The spatial distribution indicates a trend in which higher values of the intercept are found in the northern part of the region. This suggests that, accounting for spatial variations in the five variables in the model, standardized rates of LLTI still appear to be higher in the northern part of the region than in the south. The linear trend in the unemployment parameter estimate is shown in Figure 5.9 in which it can be seen that the estimate is larger in the south than in the north, suggesting that LLTI rates are more sensitive to unemployment variations in the south, which is predominantly rural. The linear trends in both the social class parameter in Figure 5.10 and the density parameter in Figure 5.11 suggest that the relationships between LLTI and social class and between LLTI and housing density both become more negative towards the more urbanized coast. Of course, these very simple trends might be disguising some interesting spatial variation that needs exploring in more detail. In order to do this, it is necessary to consider the GWR results.

5.5.4 GWR results

For the calibration of the GWR model, the spatial weighting function in Equation (5.35) was calibrated by the cross-validation technique described in Equation

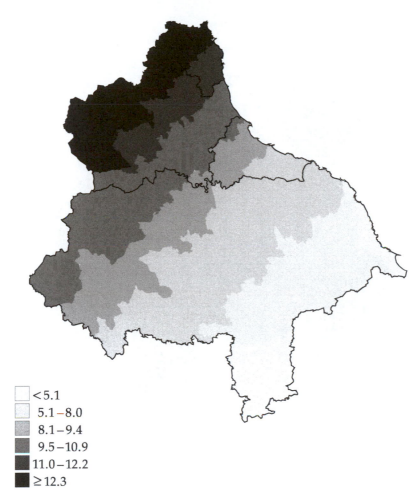

☐ <5.1
▢ 5.1–8.0
▨ 8.1–9.4
▨ 9.5–10.9
▨ 11.0–12.2
■ ≥12.3

Figure 5.8   **Linear: intercept**

(5.40). The estimated bandwidth was approximately 12 km which resulted in a weighting function that is virtually zero at a distance of approximately 19 km.[4] A plot of the CV scores against bandwidth is shown in Figure 5.12.

Local parameter estimates were then obtained by weighting the data according to this function around each point and using the estimator in Equation (5.31). The spatial distributions of the localized estimates of the intercept, unemployment, social class and density parameters are shown in Figures 5.13–5.16, respectively. The interpretation of each of the spatial estimates depicted in these figures is that it reflects a particular relationship, *ceteris paribus*, in the vicinity of that point in space.

Figure 5.13 shows the spatial variation in the estimated intercept obtained from GWR and it clearly exhibits much greater detail than the equivalent map derived from the spatial expansion method. The estimates in Figure 5.13 show the extent of

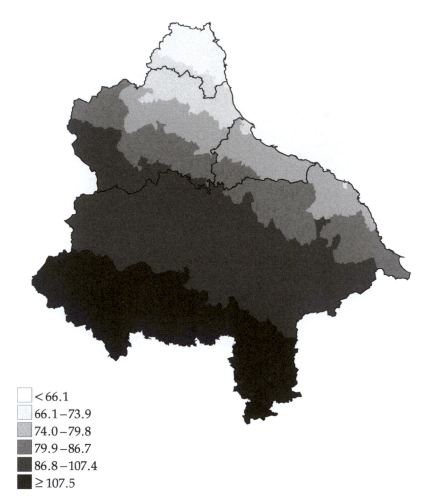

	< 66.1
	66.1 – 73.9
	74.0 – 79.8
	79.9 – 86.7
	86.8 – 107.4
	≥ 107.5

Figure 5.9   **Linear: unemployment parameter**

LLTI after the spatial variations in the explanatory variables have been taken into account. The high values that occur primarily in the industrial areas of Cleveland and east Durham suggest a raised incidence of LLTI even when the relatively high levels of unemployment and low levels of employment in professional occupations in these areas are taken into account. The Monte Carlo test of significance for the spatial variation in these estimates described above indicates that the spatial variation is significant. Presumably there are other attributes that might be added to the model that would reduce the spatial variation in the intercept. The spatial distribution in Figure 5.13 acts as a useful guide as to what these attributes should be. The model apparently still does not adequately account for the raised incidence of LLTI in the mainly industrial areas of the north-east and perhaps other employment or social factors would account for this.

The spatial variation in the unemployment parameter shown in Figure 5.14

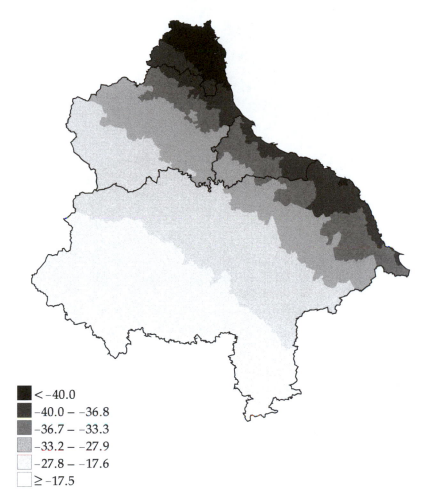

■ < −40.0
■ −40.0 – −36.8
▨ −36.7 – −33.3
▨ −33.2 – −27.9
□ −27.8 – −17.6
□ ≥ −17.5

Figure 5.10   **Linear: social class parameter**

depicts the differing effects of unemployment on LLTI across the study area. All the parameters are significantly positive but are smaller in magnitude in the urbanized wards centred on Cleveland and Tyne and Wear. Again, the spatial variation in these parameter estimates is significant. The results suggest a possible link to environmental causes of LLTI with levels of LLTI being high regardless of employment status in Cleveland which has large concentrations of chemical processing plants and, until recently, steelworks. Another possibility is that levels of LLTI are high regardless of employment status in these areas because employment is concentrated in declining heavy industries and a large proportion of the unemployed were probably formerly employed in such industries which are associated with high levels of LLTI.

   The global estimate of the social class 1 variable is significantly negative and all the spatial estimates, shown in Figure 5.15, are negative and exhibit significant

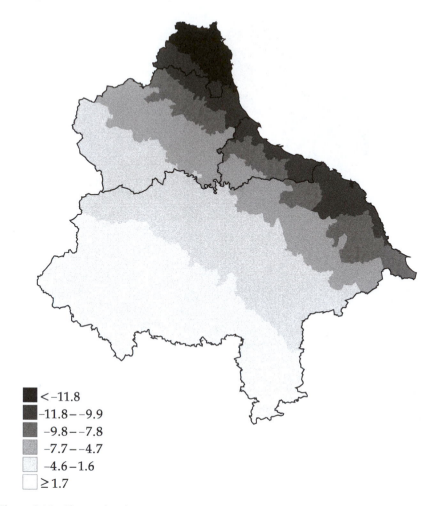

■ < −11.8
■ −11.8 − −9.9
■ −9.8 − −7.8
□ −7.7 − −4.7
□ −4.6 − 1.6
□ ≥ 1.7

Figure 5.11   **Linear: density parameter**

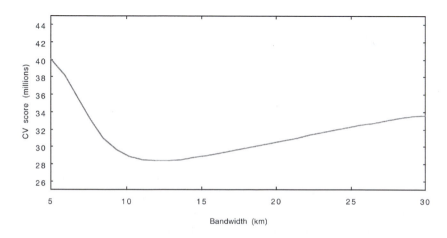

Figure 5.12   **Cross-validation scores as a function of kernel bandwidth**

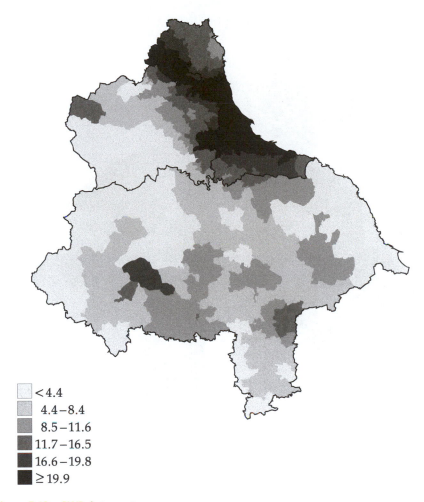

<4.4
4.4–8.4
8.5–11.6
11.7–16.5
16.6–19.8
≥19.9

Figure 5.13   **GWR: intercept**

spatial variation. The more negative estimates are concentrated along the industrial parts of Cleveland, east Durham and Tyne and Wear indicating that levels of LLTI are more sensitive to variations in social class in urban areas than in rural areas. Within urban areas LLTI is perhaps linked to manual occupations, whilst in rural areas the incidence of LLTI is more evenly distributed across types of employment.

Perhaps the best example of the improvement in understanding provided by GWR is found in the spatial pattern of the estimates for the density variable given in Figure 5.16. The global estimate for population density is significantly negative, which is somewhat counter-intuitive; we might expect that LLTI would be higher in more densely populated urban wards than in sparsely populated rural wards, *ceteris paribus*. The spatial variation of this parameter estimate indicates that the most negative parameter estimates are those for wards centred on the coalfields of east Durham. The likely explanation for this is that LLTI is closely linked to employ-

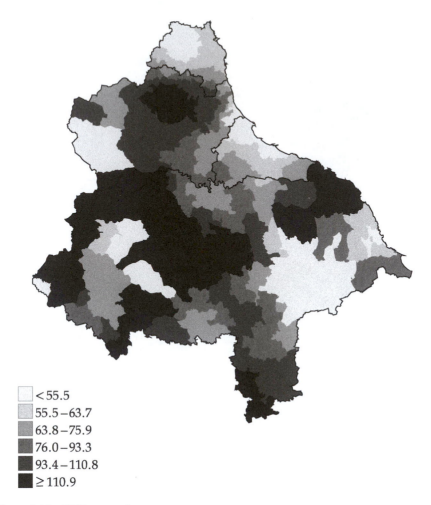

<55.5
55.5–63.7
63.8–75.9
76.0–93.3
93.4–110.8
≥ 110.9

Figure 5.14   **GWR: unemployment**

ment in coal mining (pneumoconiosis, emphysema and other respiratory diseases being particularly prevalent in miners) but that population density is low on the coalfields. The typical settlement pattern is that of scattered villages located around the coal mines. However, population density rises rapidly in the urbanized areas both immediately south and north of the coalfields where employment is less prone to LLTI. Hence, *within the locality of east Durham* it is clear that the relationship between LLTI and density is significantly negative. In the more rural parts of the study area, particularly in west Durham and North Yorkshire, the relationship is positive with *t* values in excess of 2 in many places.[5] Hence the more intuitive relationship, where LLTI increases in more densely populated areas, does exist in much of the study region but this information is completely hidden in the global estimation of the model and is only seen through GWR. The different relationships

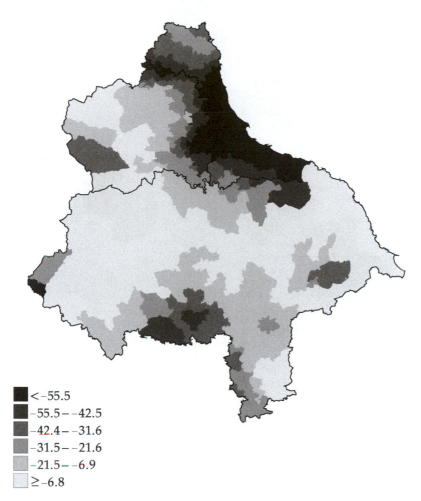

<-55.5
-55.5 - -42.5
-42.4 - -31.6
-31.5 - -21.6
-21.5 - -6.9
≥ -6.8

Figure 5.15   **GWR: social class 1**

between LLTI and population density that exist across the region and which are depicted in Figure 5.16 highlight the value of GWR as an analytical tool.

One further result from the GWR analysis is that of the spatially varying goodness-of-fit statistic, r^2, shown in Figure 5.17. These values informally depict the accuracy with which the model replicates the observed values of LLTI in the vicinity of the point for which the model is calibrated. It can be seen that there are large variations in the performance of the model across space ranging from a low of 0.23 to a high of 0.99. In particular, the model explains observed values of LLTI well in a large group of wards near the boundary between south Cleveland and the northern extremity of North Yorkshire, and also in a group of wards in the southern and westerly extremes of the study region. The model appears to replicate the observed values of LLTI less well in parts of West Yorkshire and Durham. The distribution of r^2 values in Figure 5.17 can also be used to develop the model

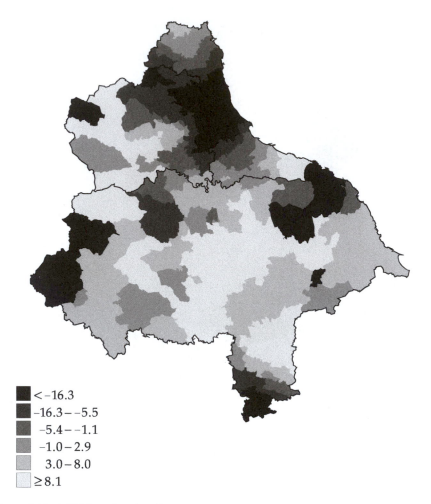

Figure 5.16   **GWR: housing density**

framework if the areas of poorer replication suggest the addition of a variable that is well represented in such areas and less well represented in areas where the model already works well. For instance, there is evidently a coalfield effect missing from the model and the low values of r^2 in West Yorkshire suggest the model still fails to account adequately for rural variations in LLTI.

Finally, it is possible to comment on the spatial variability of each parameter estimate using the results from the Monte Carlo procedure. Here, 99 random mixings of the data were undertaken and the GWR procedure was run for each data set. Thus, for each parameter, there are 100 estimates: one from the actual spatial arrangement of the data and 99 from the random arrangements of data. The variability of each set of parameter estimates is shown in the 'box and whisker' plots of Figure 5.18. The number in brackets for each parameter is the ranking of the spatial variability of each parameter estimate within the 100 sets of estimates

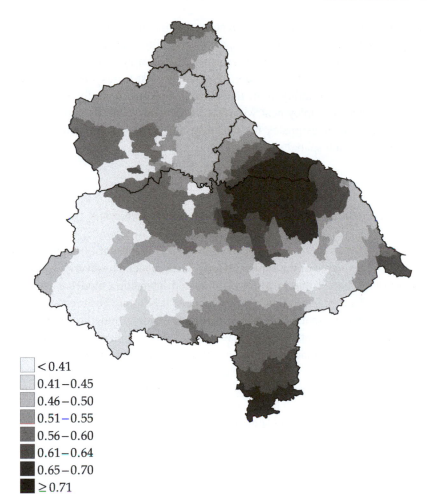

Figure 5.17   **GWR: r^2**

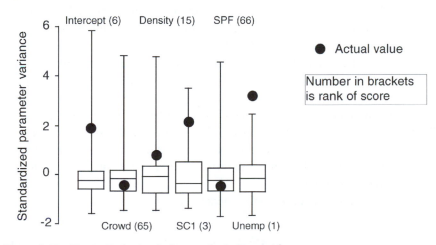

Figure 5.18   **Monte Carlo simulation results in boxplot form**

with a ranking of 1 indicating the greatest amount of spatial variability and a ranking of 100 the least amount of spatial variability. The parameter estimates obtained from the actual spatial arrangement of the data are shown as dots on the plots. It can be seen that those sets of parameter estimates that exhibit large amounts of spatial variability from the original data are those for the intercept, social class 1 and unemployment with that for density being of marginal significance. The parameter estimates for crowding and single-parent families do not exhibit any significant spatial variation.

5.6 Measuring local relationships in spatial interaction models

A substantial discussion of spatial interaction models is left until Chapter 9 where there is also an example of the application of a local model. However, it is useful to mention in this chapter that local forms of spatial interaction models have a long history and perhaps predate all other local models. It has long been recognized that global calibrations of spatial interaction models potentially hide large amounts of spatial information on interaction behaviour and that localized parameters yield much more useful information (Linneman, 1966; Greenwood and Sweetland, 1972; Gould, 1975). This was patently clear when distance-decay parameters were estimated separately for each origin in a system instead of a single global estimate being provided (see Chapter 9 for an example and further discussion).

Consider, for example, a simple global spatial interaction model of the form

$$p_{ik} = S_k^\alpha d_{ik}^\beta / \sum_j S_j^\alpha d_{ij}^\beta \qquad (5.43)$$

where p_{ik} is the probability that a person at i selects spatial alternative k; S_k is a measure of the size of alternative k; d_{ik} is the distance between i and k; and α and β are global parameters to be estimated. This model can be calibrated locally to provide separate estimates of the parameters for each origin or for each destination in the system. Often, it makes more sense to calibrate the former because the behavioural characteristics depicted by the parameter estimates tend to be a product of the origin rather than the destination. Examples include applications to the study of migration flows and store choice where we are interested in how individuals in different origins make choices. However, destination-specific models would be appropriate to investigate spatial interactions such as the choice of universities by students where the interest is on the university and how well it attracts students from a wide range of origins.

The origin-specific form of the global model in Equation (5.43) is

$$p_{ik} = S_k^{\alpha(i)} d_{ik}^{\beta(i)} / \sum_j S_j^{\alpha(i)} d_{ij}^{\beta(i)} \qquad (5.44)$$

where the parameters are now estimated separately for each origin in the system. These origin-specific parameters can be mapped to provide visual evidence of spatial variations and spatial patterns in their values (see Chapter 9 for further discussion of this issue).

5.7 Summary

The recent interest in the 'local' rather than the 'global' in quantitative geography is notable for several reasons. It belies the criticism that those adopting a quantitative approach in geography are only concerned with the search for broad generalizations and have little interest in identifying local exceptions. It links quantitative geography with the powerful visual display environments of various GIS and statistical graphics packages since the end product of local analysis is typically a spatial distribution of results which may be mapped for maximal effect. It allows spatial relationships to be explored in different ways as a catalyst to a better understanding of spatial processes and, finally, it provides an exciting framework in which to develop new statistical approaches to spatial data analysis. As Jones and Hanham (1995) suggest, local forms of spatial analysis can act as a bridge between realists and positivists.

Notes

1. Interestingly, this is what GIS generally do (see Chapter 3) and de Jong and Ottens (1977) comment on the suitability of GIS as a database for multilevel modelling.
2. Although here the spatial weighting scheme is applied to a regression model, there is no reason why such a weighting cannot be applied to many other statistics to produce local versions of them. For instance, a local, spatially varying correlation coefficient would be produced by the following weighted formula:

$$r_i = \frac{(1/n)\sum_j w_{ij}(x_j - x^*)(y_j - y^*)}{[(1/n)\sum_j w_{ij}(x_j - x^*)^2]^{1/2}[(1/n)\sum_j w_{ij}(y_j - y^*)^2]^{1/2}}$$

where r_i describes the correlation of the variables x and y around point i and w_{ij} measures the weight of point j at point i.
3. Experimental significance testing procedures are also discussed in Chapter 8.
4. For a Gaussian kernel, data at a distance of greater than $\sqrt{2}$ times the bandwidth, which in this case is approximately 12 km, are virtually zero weighted.
5. We use t values in excess of 2 here purely as an indicator of where potentially interesting relationships might occur, rather than as a test of statistical significance.

6   Point Pattern Analysis

6.1 Introduction

Spatial analysis often relies on data aggregated over a set of zones, but the patterns observed can change with the choice of zone boundaries (Openshaw, 1984). Without a clear rationale for choosing any particular set of boundaries there is no way of deciding whether observed spatial patterns are in any way meaningful. However, even if one takes this criticism of *aggregated* data analysis at face value[1] it should be noted that in most cases, aggregated data arise from a set of individual cases. The aggregation rarely takes place as an aid to spatial analysis, a much more common reason being administrative convenience. In fact it is quite possible to analyse spatial patterns of the disaggregated individual data. Usually this is done by regarding each individual case to be associated with a single point in space, and then considering the spatial distribution of these points.

Methods of these sorts are well established in a number of fields – data are available in this disaggregate form for a wide range of subjects. Point pattern analysis is particularly popular in the fields of biology (Diggle, 1983), epidemiology (Diggle et al., 1990; Bithell and Stone, 1989) and the analysis of crime patterns (Bailey and Gatrell, 1995). All of the references quoted above provide examples of the use of point pattern analysis techniques related to those described in this chapter. As with many other methods outlined in this book, several of these techniques have been developed in the last two decades, and therefore post-date the original 'quantitative revolution' in geography. It is perhaps for this reason that their existence is rarely acknowledged in critiques of quantitative approaches to geography.

As suggested above, this chapter is concerned with the analysis of point data – or at least those geographical phenomena that may be reasonably modelled as point data. This is largely a question of scale. For example, police workers will often consider household burglaries occurring in a particular city over a period of time as point data. The unit of analysis here is the household and each one is represented by a point. However, on a more local scale, two neighbouring householders in dispute over some piece of land will certainly not regard their properties as two points. Similarly, in a regional economical study a set of neighbouring towns may be viewed as zonal data but it may be more reasonable to represent all towns in the UK as points in a national study.

In terms of geographical data models, point data perhaps represent the most elementary. Each individual case in the data set is represented by a single point in space. If we analyse these data, we are essentially looking for patterns in the points; hence such activity is often termed *point pattern analysis*. As with most forms of spatial analysis, it is hoped that the patterns observed will tell us something about the underlying process that generated the points. One may wish to ask whether burglars tend to select several houses in the same neighbourhood for a 'burst' of activity, or whether they choose randomly over a wider area. Analysis of point patterns of burglaries may shed some light on this question. Alternatively, one might be interested in the locations of birds' nests. In this case, a possible hypothesis might be that some territorial birds may build nests at a distance from other nest sites – or repel other birds from building nests too near to their own. In this case, one might test for a 'dispersion' effect in a set of point data, and if this is found it may then be possible to gain some idea of the distance below which nest pairs do not occur. Again, analysis of the point patterns of nest location gives us some insight into the underlying process. These two examples demonstrate the two main phenomena that one tries to detect in point pattern analysis: namely *clustering*, where points tend to group together more than they would at random, and *dispersion*, where points tend to be spaced further apart than one would expect under a random hypothesis.

In any point pattern analysis there should be a *study area*. This is the area from which all points in the data set are drawn. Often, the data consist of *all* points in a study area, but alternatively they could be from a sample from a larger population. The latter case is discussed more fully later in the chapter. In some cases, the study area is a geometrically simple shape, such as a square or a rectangle. In other cases, however, it may take a much more complex form. In human geographical studies, for example, the study area may be contained within a set of administrative boundaries, which in many cases leads to an area having a complex and irregular shape. The choice of study area may be quite an important component of a point pattern analysis. In some cases such an area is defined intuitively – in a study of plant locations on an island, the boundary of the island itself provides an obvious study area perimeter. Likewise, if one is studying crime on a particular housing estate, then the estate boundaries should also bound the study area. In other cases there are not such obvious definitions of study area. In a biogeographic study, one may wish to look at plant locations in a 10 metre square of woodland. There is no 'natural' reason for this choice – it is perhaps the case that it is too expensive or time consuming to look at the entire woodland area, and it is hoped that this 10 metre square is in some way representative of the woodland as a whole. Another reason for the choice of a geometrically regular study region is that it may be easier to calibrate and analyse mathematical models in this framework.

Another important aspect of point pattern analysis is that of comparison. Several forms of point pattern analyses attempt to compare one set of point patterns with another. In some cases this is done between two distinct study areas – to see whether similar patterns occur in both places. On other occasions we may be looking at two or more *types* of points in the same study area. Does the presence of one species of bird's nest discourage another from building its own nest nearby? Do car thefts tend to occur in nearby places to household burglaries? These types of problem can be considered either as the analysis of several sets of point data, or as the analysis of a single set of points, with an *attribute data item*, as defined in Chapter 3. In the former case, then, we would have a database of all birds' nests in the study area, but for each nest the species of bird would be recorded, if possible. This information would be stored as a categorical attribute variable[2]. It is also worth noting that even when considering only a single set of points (without attributes), often there is still a comparative framework. Typically, here one is investigating whether the set of points comes from a 'reference' distribution. For example, in looking at disease incidence, it is often important to investigate whether cases are geographically distributed according to population density. If they are more spatially grouped than this, then one may have evidence for 'clustering' around certain places. In this case, the reference distribution is population density (or population-at-risk density) in the study area. Note that here it is not helpful to compare the disease incidence patterns with a uniform random distribution in the study area – one would expect a certain degree of grouping even without 'disease clusters', due to settlement patterns. In human geographical problems it is almost always the case that the uniform distribution is an inappropriate reference distribution.

In this chapter, methods of analysing point patterns are discussed. In the first section, exploratory approaches are considered. These are mostly graphical techniques. In the second section, different models of spatial processes are considered. It is by comparing observed patterns with these models that some insight into underlying processes may be gained. More specific methods of comparison are considered in the following sections, together with methods for comparing two or more sets of points.

6.2 Initial exploration

6.2.1 Scatterplots

Graphical approaches for all kinds of geographical data are dealt with as subjects in their own right elsewhere in this text. However, a chapter on point pattern analysis would be incomplete without some reference to this subject. Given a set of point data, it would be folly not to inspect a scatterplot of the point locations of those data. An initial visual exploration of the data should give some important hints for further data analysis. In particular, one could choose whether models based on cluster or dispersion are most appropriate. Also, inspection of a scatterplot is a rapid way of identifying any spatial outliers – that is, individual points that are located a long way apart from the main groupings. Identifying points like these can provide important insights into the outcome of an analysis – for example, do a handful of remote points exert an unreasonable amount of leverage on a statistic measuring dispersion?

A particularly important role played by graphical representations of point data is that of model specification, or *analytical strategy*. In certain cases, theory may suggest that certain types of point pattern models should be used as a basis for data analysis, but often this may not be the case, and some inspection of the point patterns themselves is needed to suggest which approaches may be helpful.

For example, consider the point pattern shown in Figure 6.1. Clearly, it would be pointless to test whether the data seen in this figure could have been generated by a random point pattern model – the regular grid arrangement of groups of points gives strong qualitative evidence against this. In fact the pattern here is so strange that one may suspect that it has occurred owing to some type of error in the recording of the data – or perhaps that some attempt at data cleaning or conversion from one file format to another has caused the pattern to occur. If this is suspected, careful consideration of the data management and data recording processes is necessary. Whether this is the case or not, this apocryphal example illustrates the importance of an initial graphical examination of the data. No amount of computer printout or tables of numbers would highlight this unusual spatial data pattern as directly as a scatterplot.

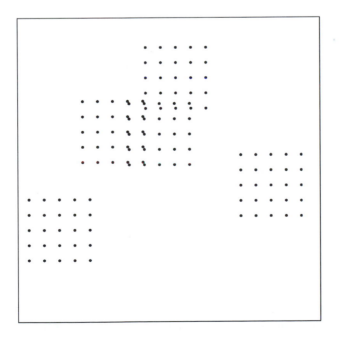

Figure 6.1   **Scatterplot of a strange point process**

6.2.2 Other exploratory plots

There are some occasions when simple scatterplots can be misleading. Consider the following example. A set of reported household burglaries is recorded by a police force over a three-month period. One particularly useful piece of information (from a spatial analyst's viewpoint) is the location of the house where the burglary occurred. It is possible to treat the location of each burgled house as an item of point data. Using a sufficiently small-scale map, or computer software to match addresses to grid references, it is possible to represent each recorded burglary as an (*x*, *y*) coordinate pair. However, in some cases, a house may be burgled several times in the three-month period. This will generate several identical (*x*, *y*) co-ordinate pairs in the data set. In a quantitative analysis this may present no difficulties, but in terms of scatterplots an obvious problem occurs. If, say, a house was burgled three times in the study period, this is represented by three points plotted in exactly the same place – which looks identical to a single point plotted in that place. Thus, in the scatterplot, multiple incidents in the same place cannot be identified, causing an understatement in the concentration of burglaries in places where repeat victimization occurs.

There are several strategies to overcome this problem. First, if the data are stored in some form of spreadsheet or database, it is possible to sort the data by the *x* coordinate, and within that by the *y* coordinate (or the other way round), and check for coincident points. If none occur, a simple scatterplot can be inspected with confidence. If there are repeat occurrences, see how often this happens, and how many data items 'collide' in each case where it does. If the maximum number of collisions is fairly small (say three or four), one can simply use different symbols on the scatterplot to indicate how many data items occur at each point.

This is illustrated in Figure 6.2. Here, household locations were recorded to 100 m grid references. This is rather imprecise for *exact* locations, and it is not uncommon for two houses having the same 100 m grid reference to be burgled in a three-month period. In fact, when treating the grid references as point data, at least one coordinate pair occurs six times in the data. Having six different symbols is rather confusing in a multiple-symbol scatterplot, and it is particularly difficult to convey ordinality in the choice of symbols. Thus, a compromise is reached here, where two different symbols are used to convey fewer than three occurrences, and three or more occurrences at a given point. From this, an area to the south-west of the study area where there is a high concentration of burglaries is apparent. The rounding to the nearest 100 m is also evident in the figure. A similar result can be seen using a *proportional symbol* or *bubble* plot – Figure 6.3. Here the *size* of the symbol reflects the number of cases occurring at a given point. Again, the area to the south-west of the study seems to have a high incidence of burglaries.

Finally, scatterplots and other forms of mapping can be enhanced by contextual information. For example, to explain why there is an area without any household burglaries in the central south area of Figures 6.2 and 6.3, it is helpful to know that this area is parkland, and so there are no houses to burgle! This kind of information

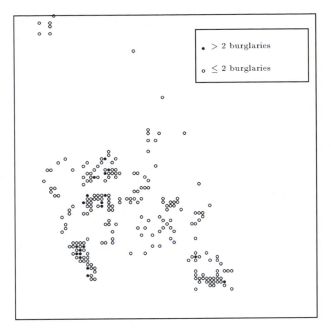

Figure 6.2   **Scatterplot of household burglary incidence**

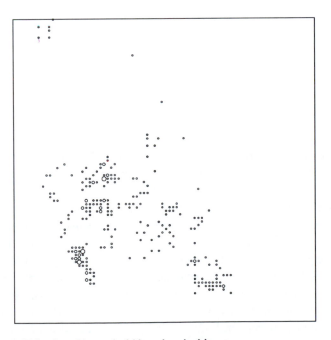

Figure 6.3   **Bubble plot of household burglary incidence**

could be marked on any of the figures shown so far in this chapter – and would again contribute in an informal way to an explanation of the process producing the point patterns observed.

6.2.3 Non-graphical approaches to point pattern exploration

In some cases, as well as plotting point patterns for an initial data examination, descriptive statistics can also give a feel for the scale and distribution of the point data. In particular, if one is comparing two or more sets of points, descriptive statistics can be used to summarize contrasts between these sets. A very basic descriptive statistic is the *mean centre*. This is the two-dimensional mean of all of the (x, y) points in a data set. If each point represented a uniform mass point in two-dimensional space, the mean centre would be the location of their centre of gravity. The formula for the mean centre is therefore given by

$$(\hat{\mu}_x, \hat{\mu}_y) = \left(\frac{\sum\limits_{i=1}^{i=n} x_i}{n}, \frac{\sum\limits_{i=1}^{i=n} y_i}{n} \right) \tag{6.1}$$

where $(\hat{\mu}_x, \hat{\mu}_y)$ is the mean centre and (x_i, y_i) are the data points. Clearly, if the points form some kind of single 'cloud' then the mean centre is a useful way of summarizing location. However, to gain some understanding of the size of this 'cloud', a second measure known as the *standard distance*, d_s, is useful. This is defined by

$$d_s^2 = \frac{\sum\limits_{i=1}^{i=n} (x_i - \hat{\mu}_x)^2 + (y_i - \hat{\mu}_y)^2}{n} \tag{6.2}$$

and is the root mean square distance of each point in the data set from the mean centre. These two statistics can be seen as two-dimensional parallels to the univariate mean and standard deviation statistics. The mean centre summarizes location, whilst the standard distance measures dispersion. However, these statistics must be treated with some caution if a point data set consists of more than one cloud, or cluster of points. For example, in the burglary data, the mean centre falls in a place where there are virtually no household burglaries (Figure 6.4). This once again illustrates the dangers of using summary statistics without graphical back-up.

Some other extremely basic statistics that also give an idea of scale and location are the minimum and maximum values of the x and y coordinates. Differences between the minimum and maximum values give some idea of the scale (or dispersion) of the process, whilst the values themselves give an impression of

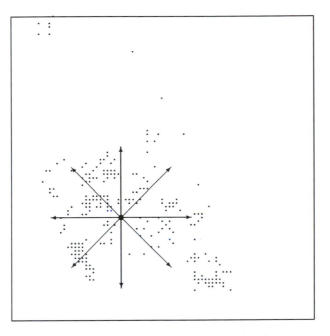

Figure 6.4   **Mean centre and standard distance for burglary incidence**

location. The main disadvantage of using these values to measure dispersion is the fact that they are easily affected by outlying data points. For example, the enclosing rectangle for the data in Figure 6.4 is made notably taller in order to accommodate a small cluster of points in the far north of the sample.

Finally in this section, as well as summarizing location and dispersion, it is also useful to summarize the degree of clustering in a set of points. The measure proposed here for this is the *mean nearest-neighbour distance*, d_m. This is simply the mean value, over all points, of the distance from one point to the closest point to it in the data set. Note that this provides additional information to measures of dispersion or location. To see this, Figure 6.5 shows two sets of points having identical mean centres and standard distances, but the points in the right hand panel have roughly double the mean nearest-neighbour distance to those on the left. As one might expect, larger mean nearest-neighbour distances suggest more regularity, whilst smaller values suggest more clustering.

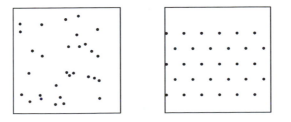

Figure 6.5   **Two very different point patterns**

It may also be useful to look at the entire distribution of nearest-neighbour distances. That is, compute the nearest-neighbour distance for each point in a data set, and consider this as a new data set. Instead of just looking at the *mean* of the distribution, it can be helpful to examine the entire distribution. One useful method here is to consider a *five-number summary* of the distribution. This – perhaps not surprisingly – is a list of five numbers which give a reasonable impression of the shape of the distribution. These numbers are the minimum, first quartile, median, third quartile and maximum of the distribution. For the burglary data, the five-number summary is given in Table 6.1.

The fact that the minimum nearest-neighbour distance is zero, as is the first quartile, shows that there is a large quantity of coincident points. The fact also that the median and third quartile are both 100 m suggests that in many other cases, the nearest neighbour is the closest 100 m grid reference to the point in question. Generally this suggests that quite a lot of clustering of points is occurring. However, the fact that in the most extreme case there is a nearest-neighbour distance of over 1500 m suggests that some outlying points are also present. This can be further investigated with a box and whisker plot (Figure 6.6).

In this case it can be seen that a small number of outlying points with uncharacteristically large nearest-neighbour distances exist. Attention should be paid to these, particularly when carrying out modelling or confirmatory analysis such as will be described in the following sections.

6.3 Modelling point patterns

In the following sections, some mathematical models for point patterns will be considered. For this, some formal definitions are required. These definitions are perhaps rather abstract in parts, and to help comprehension the linkage between the mathematical concepts and typical 'real-world' interpretations are given in Table 6.2. A *point pattern* is a data set $\mathbf{X}$ consisting of a series of two-dimensional points $\{\mathbf{x}_1, \mathbf{x}_2, \ldots\}$, where each $\mathbf{x}_i$ is a two-dimensional vector $(x_{i1}, x_{i2})^T$. These points are

Table 6.1   **Five number summary for nearest-neighbour distance of burglary data (in metres)**

Min	Qu1	Med	Qu3	Max
0	0	100	100	1526

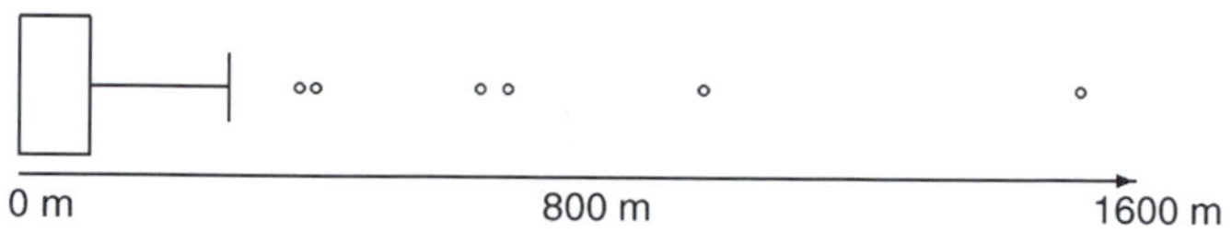

Figure 6.6   **Box and whisker plot (nearest-neighbour distances)**

Table 6.2 Mathematical concepts in point pattern analysis with some typical practical interpretations

Symbol	Generic meaning	Possible meaning in burglary example
R	Study area	Area covered by a police station
X	Event set	Locations of household burglaries in a one-month period
x$_i$	Single event i	Location of the ith burglary in a one-month period
A	Subset of study area	The extent of particular housing estate
N(A, X)	Number of events in **A**	Number of crimes on the housing estate in a one-month period
E(**A, X**)	Expected number of events in **A**	Expected number of crimes on the housing estate in a one-month period
C(x, r)	A circle of radius r around point **x**	A circle of some radius around a public house (bar) on the estate
E(**N**(**C**(**x**, r), **X**))	The expected number of events in **C(x**, r)	The expected number of crimes within a given radius of the public house (bar) on the estate in a one-month period
$\lambda(\mathbf{x}) = \lim_{r \to 0} \dfrac{E(\mathbf{N}(\mathbf{C}(\mathbf{x},\ r),\ \mathbf{X}))}{\pi r^2}$	The *intensity* of events around point **x**	The average burglary rate (in burglaries per unit area) measured in a small region centred on a point **x** in the study area

all contained within a study area **R**. In the manner of Bailey and Gatrell (1995), each of these points will be referred to as *events*, to distinguish these from other points in **R** which we may be interested in. For example, each **x**$_i$ might be the location of a household burglary, and **R** the region in Figure 6.4. Typically, we will be concerned with counting the number of events in a region, **A** say, which is a subset of **R**. In the burglary example, **A** could be a housing estate in the study area, or a circle of given radius centred on some point of interest. Denote this number of events as **N(A, X)**.

Often in a random process we will be interested in the expected number of events in a given area **A** – for example, the number of household burglaries occurring in one month on a housing estate. This can be denoted by E(**N(A, X)**). Clearly, this quantity will depend on the size and shape of **A**. A more useful quantity is the *intensity* of the process at a given point – in the burglary example, this might be the number of burglaries per square kilometre. The *mean intensity* M_1^* of a process over zone **A** measures this. M_1^* is defined by

$$M_1^* = \frac{\mathbf{N(A)}}{|\mathbf{A}|} \tag{6.3}$$

where $|\mathbf{A}|$ denotes the area of zone $\mathbf{A}$. In the burglary example, this is just the number of burglaries occurring in the housing estate divided by the area in square kilometres (or some other unit) of that estate.

Now suppose $\mathbf{x}$ is a point in $\mathbf{R}$, and that $\mathbf{C}(\mathbf{x}, r)$ is a circular sub-region of $\mathbf{R}$ centred on $\mathbf{x}$ with radius r. If we fix $\mathbf{x}$ but allow r to vary, we can divide this by the area of $\mathbf{C}(\mathbf{x}, r)$ (which is just πr^2) giving an expected rate or 'count per unit area' measure. If we let the circle shrink to a point, this expected count per unit area will tend to a limit, which is the *intensity* of the process at the point $\mathbf{x}$, denoted by $\lambda(\mathbf{x})$. Since this could be done for *any* point in $\mathbf{R}$, intensity can be thought of as a *function* or a surface defined over $\mathbf{R}$. In more formal terms,

$$\lambda(\mathbf{x}) = \lim_{r \to 0} \frac{E(\mathbf{N}(\mathbf{C}(\mathbf{x}, r), \mathbf{X}))}{\pi r^2} \tag{6.4}$$

Strictly, this limit does not have to be defined in terms of circles but could be considered as a limit of M_1^* for a general area shrinking to a point.

Note that the intensity contains information about the *absolute* rates of events occurring. One could have two processes with intensity functions $\lambda(\mathbf{x})$ and $2\lambda(\mathbf{x})$, and although in relative terms the spatial patterns of intensity would be the same, the second process would on average generate twice as many events as the first. For example, assaults might occur less frequently than vandalism, but the same areas in a neighbourhood may be 'black spots' for both of these crimes. A way to measure *relative* intensity would be to normalize $\lambda(\mathbf{x})$ by dividing by the expected number of events in the whole of $\mathbf{R}$. This gives the *probability density function* of the process, denoted by $f(\mathbf{x})$. Formally, we have

$$f(\mathbf{x}) = \frac{\lambda(\mathbf{x})}{E(\mathbf{N}(\mathbf{R}, \mathbf{X}))} \tag{6.5}$$

It is important to distinguish between probability density functions and intensity functions. On some occasions it is useful to test whether two processes have the same probability density functions, even though it may be clear that they have different intensities. For example, although assault events are rarer than vandalism, it may be interesting to test whether they have the same relative spatial distribution. This could be done by comparing the probability density functions of the two data sets. However, if one wished to test whether one crime was more prevalent than another in some area, it is intensity functions which should be compared.

The above ideas are concerned with *first-order properties* of point processes – that is, expected values associated with individual points (or areas) in $\mathbf{R}$. However, it may also be useful – particularly in terms of investigating clustering, to examine *second-order properties* of point processes. These examine the correlations or covariances between events occurring in two distinct points or regions in $\mathbf{R}$. For example, if a large number of household burglaries occur in one place, might this suggest that a large number of burglaries will also occur in another nearby place? Consider two areas in **R**, say A_1 and A_2. If these two areas were close to each other, then one would expect a clustered process to produce a positive correlation between the number of points in these two areas, if their closeness was at a similar scale to the spatial dimensions of the clustering process. To investigate these processes, we could consider the quantity $cov(N(A_1, X), N(A_2, X))$. As with the expected counts, this quantity depends heavily on the size and shape of the areas A_1 and A_2. Once again, this problem can be overcome by allowing the areas to shrink to points and taking limits of the rates. A related second-order measure of intensity is

$$M_2 = E(N(A_1, X)N(A_2, X)) \qquad (6.6)$$

This is just the expected value of the number of points in A_1 multiplied by the number of events in A_2. Going back to the burglary example, suppose now we consider two housing estates in the police beat area. The above quantity is just the average value of the number of burglaries in the first estate multiplied by the number of burglaries in the second. It is important to note the order in which this is defined – the average is taken after the multiplication takes place. In other words we consider the average of the product of the counts, not the product of the averages of the counts. The latter quantity can be expressed as

$$M_{11} = E(N(A_1, X))E(N(A_2, X)) \qquad (6.7)$$

It can be shown (see, e.g. Lee, 1997) that expressions M_2 and M_{11} are equivalent only when the number of events in A_1 and A_2 are statistically independent. In other words, $M_2 = M_{11}$ if there is no second-order interaction between counts for the respective zones A_1 and A_2. One problem with M_2 as a measure of second-order interaction is that it depends on the areas of the regions A_1 and A_2 – the larger these regions are, the greater the expected event counts and consequently the larger the value of M_2. It appears that bigger regions have greater interaction than small ones, regardless of the spatial process taking place! This difficulty can be overcome by normalizing for the areas of zones A_1 and A_2. Define a new measure M_2^* by

$$M_2^* = \frac{M_2}{|A_1||A_2|} \qquad (6.8)$$

Recall that $|A_1|$ denotes the area of a region A. Note that when the numbers of events in A_1 and A_2 are independent, a little algebraic manipulation gives the result

$$M_2^* = \frac{N(A_1)}{|A_1|} \times \frac{N(A_2)}{|A_2|} \qquad (6.9)$$

That is, M_2^* is simply the product of the first-order mean intensities (as defined in Equation (6.3)) for zones A_1 and A_2 when no second-order interaction occurs.

The above result can be used to assess the correlation between the counts for A_1 and A_2. When M_2^* exceeds the product of the first-order intensities, this suggests positive correlation, but when M_2^* is less than this product, negative correlation is implied. When the two expressions are approximately equal, no correlation is implied.

As an illustration, Table 6.3 shows the monthly burglary rates for two adjoining housing estates over a one-year period. The final column of this table gives the product of the counts for each month. Directly below the monthly counts, averages are taken, estimating the expected number of monthly counts for each housing estate. Below this the areas in square kilometres of each estate are shown; dividing the average monthly burglary counts by these areas gives the mean intensity, M_1^*, for each estate. M_2^* is computed by dividing the average product of crime counts (948.08) by the product of the areas (38.13), giving 24.86. The 'bottom line' is the comparison between M_2^* and the product of the two M_1^*. As can be seen, this product is equal to 17.77, which is less than the estimate for M_2^*. This suggests that the burglary counts for the two estates are positively correlated. Since the two estates are adjacent, this suggests that there is some degree of spatial clustering in the household burglaries.

Suppose now we need to consider the interaction between a pair of *points* in the

Table 6.3 Measuring second-order interaction for burglaries in two housing estates over a one-year period

	Burglaries		
Month	Estate 1	Estate 2	Product
January	19	40	760
February	17	19	323
March	19	37	703
April	8	22	176
May	21	45	945
June	23	43	989
July	32	55	1760
August	36	54	1944
September	28	46	1288
October	15	24	360
November	18	23	414
December	35	49	1715
Average	22.58	30.08	948.08
Area (km²)	4.1	9.3	38.13
M_1^*	5.50	3.23	
		Product	M_2^*
Comparison		17.77	24.86

study area, not a pair of *zones*. Call these points $\mathbf{x}_1$ and $\mathbf{x}_2$. In the same way as the intensity, $\lambda\mathbf{x}$, was defined, and using the **C**-notation as before, we define the second-order intensity function as

$$\gamma(\mathbf{x}_1, \mathbf{x}_2) = \lim_{r_1 \to 0, r_2 \to 0} \frac{E(\mathbf{N}(\mathbf{C}(\mathbf{x}_1, r_1), \mathbf{X})\mathbf{N}(\mathbf{C}(\mathbf{x}_2, r_2), \mathbf{X}))}{\pi^2(r_1 r_2)^2} \qquad (6.10)$$

In plainer terms, it may be verified that $\gamma(\mathbf{x}_1, \mathbf{x}_2)$ is the limit of M_2^* measured between two circular regions centred on $\mathbf{x}_1$ and $\mathbf{x}_2$, as the radii of the two circles shrink to zero. Thus, we may use $\gamma(\mathbf{x}_1, \mathbf{x}_2)$ to measure the statistical dependency between the two points. However, this is a rather abstract measure of second-order spatial dependency, and in practice the K function approach discussed later in this chapter is more frequently used.

When modelling second-order processes, certain assumptions are sometimes made. These can be seen generally as constraints on the functional form of $\gamma(\mathbf{x}_1, \mathbf{x}_2)$. Sometimes, it is only the displacement between $\mathbf{x}_1$ and $\mathbf{x}_2$ that is important. In this case, γ is only dependent on the vector difference of $\mathbf{x}_1$ and $\mathbf{x}_2$, and we may write the secondary intensity function in the form $\gamma(\mathbf{x}_1 - \mathbf{x}_2)$. A good example of this might be the locations of trees in a forest. If trees grow from seeds of older trees nearby, then the *distance* between pairs of trees may influence second-order correlations – and if there are dominant wind directions throughout the forest, then the *direction* of the line joining pairs of trees may also be important. However, within the forest, the absolute location of the tree pairs is not important. In this case, a second-order intensity function of the form $\gamma(\mathbf{x}_1 - \mathbf{x}_2)$ would be appropriate. A process which has a second-order intensity function of this form is referred to as a *stationary process*, or is said to exhibit *stationarity*.

Thus, clustering and dispersion may be modelled in many ways – either as variation in the first-order intensity, or as a consequence of the second-order intensity. It is possible for a process to have a constant first-order intensity function but form clusters due to second-order effects, and conversely it is also possible to have a process where events are independent but form clusters due to first-order effects. It is also possible that a process can be a mixture of these two: non-independent and also of non-constant first-order intensity. One of the main difficulties of point pattern analysis is the fact that for a single set of events it is not possible to distinguish between first- and second-order clustering processes.

However, this is not always a problem. Generally, a method based on second-order effects can detect clustering when it is really caused by first-order effects, and vice versa – remember, it is impossible to distinguish between the two types of process when looking at a single set of events. Thus, both kinds of test are capable of detecting clustering in general, even if they are not able to determine the *type* of clustering. The final choice of method is perhaps best guided by theoretical ideas about the underlying spatial mechanisms giving rise to the data. In this way, at least the output of the analysis may be interpreted in the context of underlying theory. In the remainder of this chapter, statistical methods for both first- and second-order clustering methods will be discussed, followed by some consideration of broader issues of point pattern analysis.

6.4 First-order intensity analysis

Perhaps the most obvious way of estimating a first-order intensity function is to divide the study area into grid squares, count the number of events in each square, and then divide these counts by the areas of the squares. One data set which can be studied in this way is the Redwood data (Diggle, 1983), shown in Figure 6.7.

Essentially, the point events here are the locations of redwood trees in a 10×10 m square of Canadian forest. Arguably, the pattern seen here could have been produced from either a first- or a second-order process – the model of seed propagation discussed in the previous section could be applicable here, but intensity might also be governed by soil fertility, suggesting a first-order process. Here, the data will therefore be examined in both frameworks.

Applying the point-counting method here, using a 2 m grid square, we obtain the pattern in Table 6.4. To compute the estimated intensity function $\lambda(\mathbf{x})$, one simply finds which grid square $\mathbf{x}$ lies in, and then uses the value of density computed for that square. Clearly, this method is very sensitive to the size of the grid squares being used.

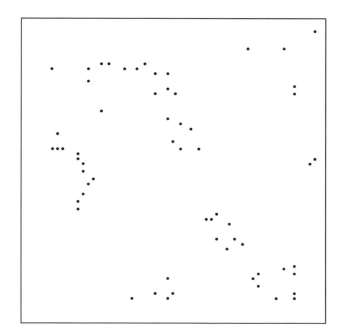

Figure 6.7 **Redwood locations**

Table 6.4 **2 m grid counts for Redwood sapling data**

2	2	2	0	6
1	0	0	8	3
3	6	3	0	4
5	2	5	0	1
1	1	6	1	0

Having computed an estimate of the intensity function, the next problem is perhaps to test for clustering. One way of doing this is to consider departures from an 'unclustered' model. In this case, an appropriate reference point might be a model of *complete spatial randomness* or CSR. A CSR process is characterized by a constant first-order density function over the study area, and by independence events. In mathematical terms, we have $\lambda(\mathbf{x}) = \lambda$ and $cov(\mathbf{N}(\mathbf{x}, \mathbf{A}_1), \mathbf{N}(\mathbf{x}, \mathbf{A}_2)) = 0$ providing $\mathbf{A}_1$ and $\mathbf{A}_2$ are non-overlapping regions. In this section, the first-order property $\lambda(\mathbf{x}) = \lambda$ is of most importance.

If we had CSR here, what event counts would we expect to see in the grid squares? More specifically, if $\mathbf{G} = \{g_1, g_2, \ldots\}$ is the set of counts in each of the grid squares, what could we say about their probability distribution? First, under CSR, since no grid square overlaps another, the second condition allows us to assume that the counts are *independently* distributed. Secondly, since the intensity is constant throughout the study area, we can also assume that the counts are *identically* distributed. Finally, and requiring a little more mathematics, it can be shown that each grid count is *Poisson* distributed, with a mean equal to $\lambda|A_g|$ where $|A_g|$ is the area of a grid square. The core of this argument is that a point appearing in a particular grid square may be treated as a 'rare event'. A simple way to assess whether we have CSR, therefore, is to see whether the point counts $\mathbf{G}$ appear to follow a Poisson distribution. This can be done using a χ^2 goodness-of-fit test. Note that, in effect, an infinite number of tests *could* be performed, as one could choose any size for the grid squares. However, certain practical issues would constrain this choice.

Another, perhaps more informative, approach is based on an interesting property of the Poisson distribution: for any Poisson distribution, the mean is equal to the variance. Thus, the ratio of the mean to the variance should be approximately one, if the null hypothesis of CSR holds. This leads to the test statistic

$$I = \sum_{i=1}^{n}(g_i - \overline{g})^2/(n-1)\overline{g} \tag{6.11}$$

which is based on the ratio of the sample variance to the mean. Here, n is the total number of grid squares covering the study region. This index is called the *index of dispersion*. In the case of CSR, both the sample variance and the sample mean are estimates of the same quantity, and the expected value of I is 1. In fact, under CSR we have that $(n-1)I$ is approximately distributed as χ^2_{n-1}. This approximation holds well provided $n > 6$ and $\lambda|A_g| > 1$. Note that values of I very much less than 1 suggest dispersion, or regularity in point patterns (as they suggest the variability in grid counts is much less than under CSR), and values much larger than 1 suggest clustering.

As an example, consider the Redwood sapling data. Dividing the study area into 2 m grid squares, there are then 25 observed counts in all. These are set out, in spatial order, in Table 6.4. The mean of these counts, which is an estimate of $\lambda|A_g|$, is 2.48. This exceeds 1, and n is 25, so the χ^2 test is valid. In this case $I = 2.22$, which should be multiplied by 24 and compared against the χ^2_{24} distribution to test CSR as a null hypothesis. Since 24×2.22 is 53.28, and the 99% point of the χ^2_{24} distribution is 42.98, we may reject a null hypothesis of CSR at the 1% level. Since this is an upper tail test, we reject the null hypothesis in the direction of clustering. It would seem that Redwood saplings group together.

It is worth noting that the size of grid squares used in this technique is an arbitrary choice, provided it satisfies the conditions set out earlier. This could be regarded as a disadvantage of the method – it is possible that one choice of grid square may fail to detect a deviation from CSR when another may not. To see how this effect may vary, consider the result when the index of dispersion is computed using several different grid sizes for the Redwood data (Table 6.5).

Clearly, not all of these combinations would be valid for significance testing, and there would be obvious problems with using a table like this to 'hunt' for significant results. However, some idea of the scales which seem best for picking up clustering can be seen from this table. It is also worth noting that at all scales, $I > 1$, so we can be fairly confident in the diagnosis that some form of clustering is present.

6.4.1   Kernel density estimates

The grid square point count approach has two obvious disadvantages. First, as outlined above, the choice of grid size can be arbitrary. In the very worst cases, one might deliberately 'hunt' for a grid size that gives the desired result of a significance test for CSR. The second disadvantage is the discrete nature of the estimate for $\lambda(\mathbf{x})$. Each grid square has its own estimate, based on the number of points observed in itself, but this estimate undergoes a 'quantum leap' each time a grid square boundary is crossed. In this section, *continuous* estimates of $\lambda(\mathbf{x})$ will

Table 6.5   **Index of dispersion (I) for different grid sizes (Redwood data)**

Grid size	5.00	3.33	2.50	2.00	1.67	1.43	1.25	1.11	1.0
I	1.65	2.84	2.40	2.22	1.88	2.23	2.34	1.77	1.95
n	4	9	16	25	36	49	64	81	100
$\lambda|A_g|$	15.5	6.89	3.87	2.48	1.72	1.27	0.97	0.77	0.62

be considered. Essentially, these are spatial extensions of the methods discussed in Section 4.5.

One basic way of estimating the value of $\lambda(\mathbf{x})$ at a given point $\mathbf{x}$ would be to draw a circle around $\mathbf{x}$ of a given radius, and then count the number of points in it. Dividing this point count by the area of the circle would then give an estimate of the intensity. This approach is sometimes called the naive estimator, defined as

$$\lambda_n(\mathbf{x}) = \frac{N(C(\mathbf{x},\, r))}{\pi r^2} \tag{6.12}$$

It has the advantage over the grid count method that there is no directional bias, and it is focused on the point $\mathbf{x}$ that is of interest. However, some major problems are not addressed. The first of these is that the search radius, r, is arbitrary, in much the same way as the size of grid squares in the previous approach. Secondly, the method also exhibits some degree of discontinuity. As $\mathbf{x}$ moves around the study area, events will move in and out of the search circle. As they do so, the estimate of $\lambda(\mathbf{x})$ will make discontinuous jumps.

One way of looking at this is to note that events are either inside the circle or outside it. There is no 'partial' or 'weighted' counting, where an event might contribute a fractional component to the numerator of Equation (6.12). However, if one down-weighted the contribution of more distant event points when estimating the intensity at point $\mathbf{x}$, the discontinuity problem might be avoided. This can be achieved by centring a small hump over each event point. The height of this hump at $\mathbf{x}$ represents the contribution towards the point count estimate used to estimate $\lambda(\mathbf{x})$. For event points a long way from $\mathbf{x}$, this contribution is very small, but for event points close to $\mathbf{x}$ it is much larger. Mathematically, the hump may be described as a two-dimensional probability density function. Suppose we have a general distribution $k(\mathbf{x})$, with a single mode at $\mathbf{x} = 0$; then if $\mathbf{x}_p$ is an event point the related probability distribution $1/h^2 k(\mathbf{x} - \mathbf{x}_p/h)$ represents the hump. Here, the shape of the function $k(\mathbf{x})$ determines the shape of the hump – for some examples see Chapter 3. Also, the parameter h (the *bandwidth*) controls the spread of the hump – it is very similar to r in the naive estimator. Large values of h give very widely spread humps. Since these humps are actually probability density functions, they are already intensity measures, with a total intensity of one over the entire study area. Thus, if each of these hump functions is summed at a given point, this will give an estimate of $\lambda(\mathbf{x})$. If we refer to this estimate, the *kernel intensity estimate* of $\lambda(\mathbf{x})$, as $\hat{\lambda}_k(\mathbf{x})$, then more formally we can say

$$\hat{\lambda}_k(\mathbf{x}) = \sum_{i=1}^{i=n} \frac{1}{h^2} k\left(\frac{\mathbf{x} - \mathbf{x}_i}{h}\right) \tag{6.13}$$

Note that as long as $k(\mathbf{x})$, the *kernel function* is continuous, so is $\hat{\lambda}(\mathbf{x})_k$. Note also that dividing the kernel intensity estimate by n yields the kernel probability density estimate.

A kernel-based intensity estimate of the Redwood distribution is shown in Figure 6.8, with $h = 1$ m. The darker areas correspond to larger values of intensity. Here, h is chosen intuitively. However, as suggested earlier, choice of h is quite a complicated matter. Typically, too large h values tend to smooth out interesting effects (such as local maxima in the distribution), whilst too small h values tend to produce 'spikey' intensity functions, with spikes centred on the observed data points. Some form of 'optimal' choice for h would be useful. Intuitively, one would expect this choice to lie somewhere between oversmoothing and 'spikiness'. A number of possible methods have been suggested. Many of these are chosen on the basis of the MISE (Mean Integrated Squared Error), a measure of discrepancy between the kernel probability density estimate ($\hat{f}$, say) and true probability density estimate f:

$$\text{MISE} = \text{E}\left\{ \int [\hat{f}(\mathbf{x}) - f(\mathbf{x})]^2 \, d\mathbf{x} \right\} \tag{6.14}$$

This quantity is approximately optimized by Bowman and Azzelini (1997) as

$$h_{\text{opt}} = \left[\frac{\{\int (k(\mathbf{z}))^2 \, d\mathbf{z}\}^2}{n[\text{var}(k)]^2 \int (\nabla^2 \cdot f(\mathbf{z}))^2 \, d\mathbf{z}} \right]^{1/6} \tag{6.15}$$

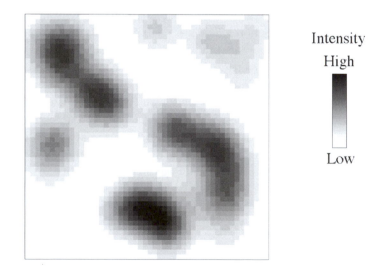

Figure 6.8   **Intensity plot for Redwood data**

for bivariate kernels which are the product of two univariate kernels. Unfortunately, this rather complex expression requires f to be known – which is rather unhelpful as one is attempting to estimate f! However, one can start by considering the situation when f is a normal distribution, in which case we have (for a two-dimensional smoother)

$$h_{\text{opt}} = \left[\frac{2}{3n}\right]^{1/4} \sigma \qquad (6.16)$$

where σ, the standard deviation of f, could be estimated from the sample. For example, a standard distance (as defined earlier in this chapter) could be used. This tends to oversmooth, as a normal distribution is one of the smoothest possible distributions. However, there are arguments in favour of this conservative approach: any maxima observed in estimated intensity or density curves are more likely to be genuine rather than an artefact of undersmoothing. Terrell (1990) took this approach a stage further, by choosing h to be the largest possible smoothing parameter for *any* distribution with variance σ^2.

Other forms of selection are possible – see for example Brunsdon (1995a) or Scott (1992). Alternative approaches allow for *local h* values, so that the intensity estimate becomes

$$\hat{\lambda}_k(\mathbf{x}) = \sum_{i=1}^{i=n} \frac{1}{h_i^2} k\left(\frac{\mathbf{x} - \mathbf{x}_i}{h_i}\right) \qquad (6.17)$$

This is useful when sample points vary greatly in density – for example, locations of households in a study area covering both urban and rural regions. In the more dense areas, a smaller h could identify local patterns in reasonable detail, but in the sparser areas using this same bandwidth might result in a series of spikes around the sample points. Variable h techniques are discussed in detail in Silverman (1986) and Brunsdon (1995a).

6.5 Second-order intensity analysis

In this section, the calibration of second-order processes will be considered. Recall that these processes can be characterized by a *second-order intensity function* $\gamma(\mathbf{x}_1, \mathbf{x}_2)$, a measure of the dependency between the two points $\mathbf{x}_1$ and $\mathbf{x}_2$. Also, for an isotropic, stationary process the second-order intensity function depends solely on the distance between the two points. In this case, an alternative (and perhaps more tangible) way of characterizing the second-order properties of the process is the use of K functions (Ripley, 1981). A K function of a distance d is defined by

$$K(d) = \frac{E\mathbf{N}(\mathbf{C}(\mathbf{x}, d))}{\overline{\lambda}} \tag{6.18}$$

where $\mathbf{C}$ is the 'circular region' operator, $\mathbf{N}$ is the 'point-counting' operator, and $\overline{\lambda}$ is the 'average intensity' of the process – that is, the mean number of events per unit area. In plain terms, the K function for a distance d is the average number of events found in a circle of radius d around an event, divided by the mean intensity of the process. The mean intensity of the process, $\overline{\lambda}$, is simply the number of events divided by the study area. It is important to note that the term 'event' is used twice in this definition – the numerator here is not the average number of events found in a circle of radius d *anywhere* in the study area, only in circles centred on other events. In the burglaries example it is the average number of burglaries found in a circle of radius d centred on another burglary. Clearly, this is a second-order property, counting distance relationships between *pairs* of events. Like γ, it is a way of measuring the likelihood of further events occurring near to an initial event. The numerator in Equation (6.18) may be interpreted as a standardization for sample size, n. Clearly, if there are more points in the sample one expects to find more events in circles drawn around event points, but here we wish to consider a characteristic of the underlying spatial process which is independent of n.

The advantage of the K function is that it may be easily estimated. The average intensity is just the number of events in a sample, n, divided by the area of the study region, $|A|$. For a given distance d, one can count the number of pairs of events that are less than d apart, and divide this by n, to obtain a sample mean estimate of $\mathbf{N}(\mathbf{C}(\mathbf{x}, d))$. For example, in Figure 6.9 the locations of Redwood saplings are shown, together with rings of radius 0.4 m around each of them. Counting the number of points falling into a ring gives the number of saplings lying within 0.4 m of the corresponding sapling. Averaging this number for all of the rings gives an estimate of $E[\mathbf{N}(\mathbf{C}(\mathbf{x}, d))]$ for $d = 0.4$ m. Dividing this by the average intensity gives a basic estimate for $K(d)$, where $d = 0.4$ m. In Table 6.6, the counts, the counts of saplings within 0.4 m for each of the 62 saplings are shown. The average value for these counts is 0.935, so that the average number of saplings within 0.4 m of another sapling is just under one. The average intensity here is 62 saplings divided by the area of the sample region, which is 10 m $\times$ 10 m, or 100 m^2. This gives 0.62 saplings per square metre. Dividing the previous result by this figure gives

$$\frac{0.935}{0.62} = 1.508$$

which is an estimate for $K(0.4)$, when distances are measured in metres. That is, the average number of saplings one expects to find within 0.4 m of another sapling, estimated earlier as 0.935, can be expressed as 1.508 times the average intensity of 0.62 saplings per square metre. Similar estimates can be made for other d values,

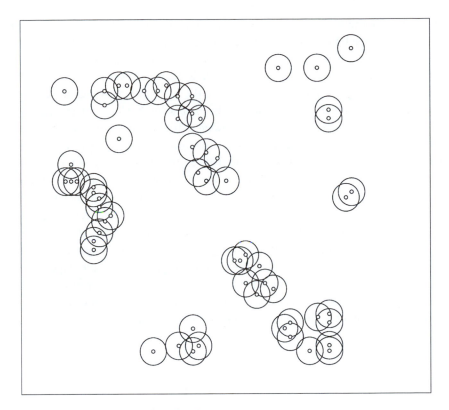

Figure 6.9 **Estimating the *K* function for *d* = 0.4 *n***

Table 6.6 **Counts of other saplings within 0.4 m for each sapling**

0	1	2	1	0	0	1	0	0	0
2	3	2	1	1	1	1	0	0	1
0	1	1	2	2	2	0	0	1	1
0	2	0	0	1	1	2	2	1	2
2	0	1	2	1	1	0	0	1	2
0	2	1	0	1	2	1	1	0	1
1	0								

by using different circles of different radii in Figure 6.9. If we call an estimate of this kind $\hat{K}(d)$ then we have

$$\hat{K}(d) = \frac{|A|}{n^2} \sum_{i=1}^{i=n} \mathbf{N}(\mathbf{C}(\mathbf{x}_i, d)) \tag{6.19}$$

If it is known that the study area does not contain the entire spatial process under study, it is also important to consider edge effects. For events close to the edge of the study area, there is a chance that other events just outside the study area will not be represented in $\hat{K}(d)$, causing an underestimate of $K(d)$. There are two relatively simple ways to overcome this difficulty. First, if d is relatively small compared with the study area, one could exclude observed events lying closer than d to the boundary of the study area from the summation in Equation (6.19). However, these points *are* still counted in the expression $\mathbf{N}(\mathbf{C}(\mathbf{x}_i, d))$, so there is no under-representation of point counts. However, if sample sizes are small, or d values of interest are relatively large, the above method may discard an unacceptable amount of data. In this case, an alternative strategy is to adjust the counts of events centred around events less than a distance d from the boundary. Essentially, the adjustment is made on the basis that, since not all of $\mathbf{C}(\mathbf{x}_i, d)$ lies inside the study area, the count process could miss out certain events, resulting in an undercount. If we assume fairly constant intensity throughout $\mathbf{C}(\mathbf{x}_i, d)$, and assume that n_i events fall within $\mathbf{C}(\mathbf{x}_i, d)$, including those outside the study region, then if α_i is the proportion of $\mathbf{C}(\mathbf{x}_i, d)$ lying in the study region, one would expect to see $\alpha_i n_i$ of these events in the sample. Since $0 \leqslant \alpha \leqslant 1$ this is an underestimate as one would expect. However, a reasonable estimate of the 'true' number of events could be obtained by dividing the observed number by α_i. In this case, we modify our definition of $\hat{K}$ as below:

$$\hat{K}(d) = \frac{|A|}{n^2} \sum_{i=1}^{i=n} \frac{\mathbf{N}(\mathbf{C}(\mathbf{x}_i, d))}{\alpha_i} \tag{6.20}$$

Note that here $\mathbf{N}$ denotes a count of events appearing *in the sample*, and that if $\mathbf{x}_i$ is further than a distance d from the boundary, then $\alpha_i = 1$.

One useful application of this technique is in testing for clustering. This is done by comparing the estimated K function for the sample data with the theoretical K function for a process with uniform intensity. If a process has uniform intensity of, say, λ, then one would expect to find $\lambda \pi d^2$ events within a distance d of an event – this is just the intensity multiplied by the area of the study region within a distance d of λ. Since the average intensity of the process is just λ, the theoretical K function is then πd^2. Thus, comparing $\hat{K}(d)$ with πd^2 gives a useful check for clustering or dispersion. Generally, if $\hat{K}(d)$ is greater than πd^2 then events tend to have more other events near them than one would expect at random – suggesting clustering – whilst if $\hat{K}(d)$ is less than πd^2 then events tend to have fewer events near to them than one would expect at random – suggesting dispersion.

For example, using the Redwood sapling data, we have already estimated that $K(0.4) = 1.508$. Hypothesizing that the Redwoods were uniformly distributed over the study area, we would expect the theoretical K function value to be πd^2, or approximately $3.1415 \times 0.4 \times 0.4 = 0.503$. Thus, the observed K function value is about three times that expected under a hypothesis of uniformity. This suggests that some degree of clustering is occurring at this spatial scale.

The comparison can be made easier if one plots the difference between $\hat{K}(d)$ and πd^2, or some other expression that is zero when the two functions are equal, and whose sign indicates which of the two functions is greatest. Examples include

$$\hat{L}(d) = \sqrt{\frac{\hat{K}(d)}{\pi}} - d \tag{6.21}$$

$$\hat{l}(d) = \frac{1}{2} \log\left(\frac{\hat{K}(d)}{\pi}\right) - \log(d) \tag{6.22}$$

$$\hat{\Delta}(d) = \hat{K}(d) - \pi d^2 \tag{6.23}$$

For the Redwood data, a $\hat{K}(d)$ is computed using formula (6.20), giving the left hand graph in Figure 6.10. This is compared with a reference K function for a uniform distribution. At first glance, the two curves appear very similar, but close inspection shows that $\hat{K}$ is generally larger than the reference curve. The difference does not show up clearly since the scale of the graph has to be large enough to accommodate high values of the K functions, but the major differences are at the lower values. In situations like this, the other comparison functions produce more illuminating graphs. On the right of Figure 6.10, the comparison function $\hat{l}(d)$ defined in Equation (6.22) is plotted. From this, the higher values of the K function below a distance of about 3 m become apparent. This suggests that Redwood saplings tend to cluster in the sense that there are more pairs of saplings within 3 m of each other than one might expect if the saplings were randomly distributed.

However, it should be noted that $\hat{l}(d)$ is a *sample* function, and therefore subject to random error. It would be useful to obtain some kind of confidence interval around $\hat{l}(d)$ for each d. One possible approach to this is to use *bootstrap* or *Monte Carlo* methods (Efron, 1982) – see also Chapter 8. To model the sampling of $\hat{l}(d)$ theoretically, one would need to know the true spatial distribution of the point pattern. However, this sampling distribution can be approximately simulated by sampling n points with replacement from the data set. Repeating this exercise a large number of times for a given d, and finding the mean and upper and lower 5% quantiles for the repeated estimates of $\hat{l}(d)$, gives a set of estimates for $l(d)$, with approximate upper and lower 5% confidence limits. This is shown in Figure 6.11 for 1000 simulations. This figure seems to bear out the conclusions reached earlier. Note that the confidence bands are *pointwise*, so that there are some curves falling within the confidence bands which might not fall within a 'confidence set of curves'[3] for $l(d)$.

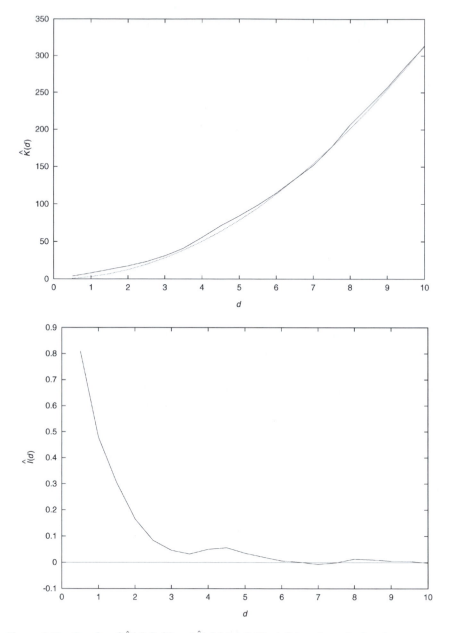

Figure 6.10 Graphs of $\hat{K}(d)$ (left) and $\hat{l}(d)$ (right). The left hand graph also shows the reference curve πd^2 (dotted)

6.6 Comparing distributions

The final important task in point pattern analysis to be considered is the comparison of point patterns. Given a pair of point patterns, are they likely to have come from the same distribution? This is a particularly important question in epidemiology,

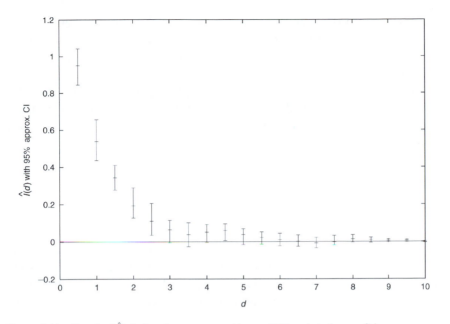

Figure 6.11   Graph of $\hat{l}(d)$ showing upper and lower 95% pointwise confidence bands

when one frequently wishes to test whether the spatial distribution of a particular disease is different from the spatial distribution of the population at risk. One approach here is a *case-control* study. A set of matched 'controls' – people who have not contracted an illness under investigation – are chosen from the population, and their distribution is compared with that of the 'cases' – people who have contracted the illness. A similar approach could be applied to crime data – for example, to compare the distributions of burgled households with those that have not been burgled over a given time period. There are also many applications in physical geography – for example, to compare the spatial distributions of different species of plants or sinkholes in a karst region.

As before, it is possible to approach the problem of distribution comparison in terms of either first- or second-order properties. Taking two sets of events, both in the same study area, one can compare kernel density estimates (first order) or estimated K functions (second-order). First, the approach for the kernel density estimate will be considered.

6.6.1   Comparing kernel densities

Clearly, if there are two sets of point events, then two kernel density estimates can be computed. Call these $\hat{f}(\mathbf{x})$ and $\hat{g}(\mathbf{x})$. It is intended to determine whether these are both sample estimates of the same distribution, and, if not, then what the geographical patterns are in the difference. For example, it is much more helpful to know *where* the distributions for cases and controls differ (and whether cases are more prevalent than controls or less so) than simply to know that the distributions differ. An intuitive comparison technique would be to plot the function $\hat{f}(\mathbf{x}) - \hat{g}(\mathbf{x})$.

Since both of these estimates are subject to sampling error, it seems reasonable to compute confidence intervals (or at least standard errors) for $\hat{f}(\mathbf{x})$ and $\hat{g}(\mathbf{x})$, and consequently for the difference between them. It can be shown (see Bowman and Azzelini, 1997) that the variance of $\hat{f}(\mathbf{x})$ is approximately

$$\frac{1}{h^2 n} f(\mathbf{x}) \left[\int (k(\mathbf{x}))^2 \, d\mathbf{x} \right]^2 \tag{6.24}$$

where k is the kernel function (assumed to be the product of two univariate kernels). This allows us to compute the standard error of $\hat{f}(\mathbf{x})$ (and similarly $\hat{g}(\mathbf{x})$), but requires that the true f (or g) is known. Clearly, this is not particularly helpful. However, a more useful related result is found by taking the square root of the density estimate:

$$\text{var}\left\{ \sqrt{\hat{f}(\mathbf{x})} \right\} \approx \frac{1}{4} \frac{1}{nh^2} \left[\int (k(\mathbf{x}))^2 \, d\mathbf{x} \right]^2 \tag{6.25}$$

This no longer depends on the true f or g. Thus, it seems more appropriate to consider the difference between $\sqrt{\hat{f}(\mathbf{x})}$ and $\sqrt{\hat{g}(\mathbf{x})}$. Note that provided $\hat{f}$ and $\hat{g}$ are both estimated using the same kernel bandwidth, the expression

$$\sqrt{\hat{f}(\mathbf{x})} - \sqrt{\hat{g}(\mathbf{x})} \tag{6.26}$$

will have mean zero, if f and g are identical. We already know that this expression has constant variance with respect to $\mathbf{x}$, providing the samples for f and g are independent. Dividing expression (6.26) by the square root of this variance we obtain a quantity which, if f and g are identical, has zero mean and unit variance for all $\mathbf{x}$ in the study area. Thus, let us define the function $\delta(\mathbf{x})$ by

$$\delta(\mathbf{x}) = \frac{\sqrt{\hat{f}(\mathbf{x})} - \sqrt{\hat{g}(\mathbf{x})}}{\sqrt{[\int (k(\mathbf{x}))^2 \, d\mathbf{x}]^2 / [2(n_f^{-1} + n_g^{-1})h^2]}} \tag{6.27}$$

where n_f and n_g are the respective sample sizes for the two groups. A plot of $\delta(\mathbf{x})$ is a useful tool for comparing f and g. Drawing contours around $\delta(\mathbf{x}) = 2$ and

$\delta(\mathbf{x}) = -2$ is a useful method of identifying areas where f and g are 'notably' different.

Although this provides a useful method for comparing distributions, this should not be considered as a significance test. First, one is carrying out an infinite number of tests (i.e. one for every value of $\mathbf{x}$ in the study region); secondly, $\delta(\mathbf{x})$ values for the $\mathbf{x}$ close to each other will not be independent; thirdly, see the critique of significance tests in general in Chapter 8.

As an example, the burglary data discussed earlier in the chapter are revisited. As well as the location of each burglary, the method of entry used by the burglar was also recorded. In particular, it was recorded whether or not 'lock drilling'[4] was used. Thus, there are two event sets, for 'drillers' and 'non-drillers'. It is useful to know whether there is any difference in the spatial distribution of the two sets.

By estimating f (for drillers) and g (for non-drillers) with a Gaussian kernel with a bandwidth of 1.5 km, a contour plot of $\delta(\mathbf{x}) = \pm 2$ is produced (Figure 6.12). From this, two main areas are identified – one to the west of the study region (where $\delta(\mathbf{x}) > 2$) corresponding to a dominance of drillers, and another to the east of the study area (where $\delta(\mathbf{x}) < -2$) where non-drillers dominate. A small area to the north of the map where drillers dominate is also present. Given the small size of this latter area, it is possible that this may be a spurious result, particularly since the area is also close to the edge of the study region where relatively few observations exist. For the two larger areas, it is interesting to note that the area to the west is a socially disadvantaged area, and the area to the east is relatively affluent – perhaps there is a greater presence of 'drillable' locks in the former area.

6.6.2 Comparing K functions

Second-order properties may also be considered when comparing distributions of spatial point patterns. In particular, one could use the techniques outlined in Section 6.5 to estimate K functions for a pair of event sets, and compare these. This approach is very similar to the simple test of clustering using K functions. Here, instead of comparing an estimated K function with a theoretical K function, two estimated K functions are compared. Indeed, the comparison functions suggested by Equations (6.22) and (6.23) can be modified to compare two estimated K functions:

$$\tilde{l}(d) = \log(\hat{K}_1(d)) - \log(\hat{K}_2(d)) \qquad (6.28)$$

$$\tilde{\Delta}(d) = \hat{K}_1(d) - \hat{K}_2(d) \qquad (6.29)$$

Although either of these functions may be useful, we would need to know the expected sampling distributions of $\tilde{l}(d)$ or $\tilde{\Delta}(d)$ when the two true K functions were identical in order to interpret them meaningfully. Unfortunately, it is not

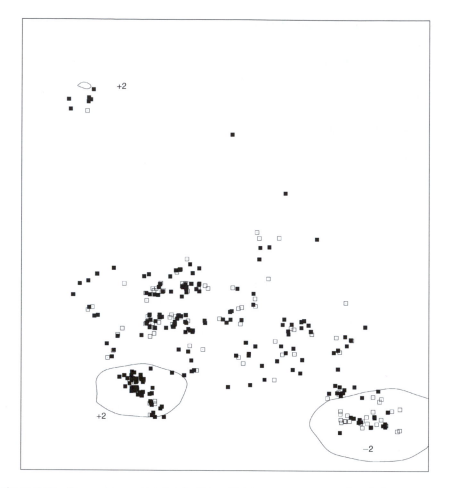

Figure 6.12 **Comparing spatial distributions. Light squares represent burglaries where locks were not drilled, dark squares represent burglaries where locks were drilled. Contours for $\delta(x) = \pm 2$ are shown as lines**

possible to derive these distributions analytically. However, a simulation-based approach may be adopted. Consider the two point data sets as a single, merged data set with a binary labelling variable. Assuming there is no difference between the two spatial distributions, any permutation of labelling amongst the event points is equally likely. By drawing several random permutations of labels and computing $\tilde{l}(d)$ or $\tilde{\Delta}(d)$, the distributions of the two functions can be approximated. Another simulation-based approach is to compute approximate confidence limits for $l(d)$ or $\Delta(d)$ using the bootstrap method described in Chapter 8.

Shown in Figure 6.13 are bootstrap estimates (including confidence limits) for $l(d)$ for the burglary data split into drillers and non-drillers. These suggest that there are no strong differences in the second-order properties of the two processes. Although the distributions are clustered in different places, the geographical scale of clustering is similar.

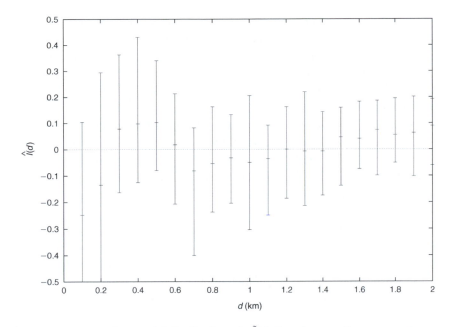

Figure 6.13   **Comparing spatial distributions via $\tilde{I}(d)$. Error bars are the upper and lower 95% CIs**

Other kinds of analyses are also possible. An interesting possibility is the *cross-K function*, $K_{12}(d)$, defined as the expected number of group 2 events within a radius d of group 1, divided by λ_1, the mean intensity of group 1. Clearly, K_{21} is defined symmetrically. Note that these functions measure the tendency of one type of event to cluster around, or be repelled from, clusters of another type of event. In the burglary example, 'repulsion' could be observed if one group of burglars (the drillers) tended to operate in different neighbourhoods to the other group (the non-drillers). It can be argued (Bailey and Gatrell, 1995) that if both groups are generated by the same spatial process, one would expect the cross-K function to be πd^2. It is possible to estimate the cross-K function in exactly the same way as it is to estimate an ordinary K function. One can then estimate the differences between the estimated cross-K function and πd^2 using a comparison function such as $\hat{I}(d)$, $\hat{L}(d)$ or $\hat{\Delta}(d)$.

6.6.3   *Comparing a point pattern with a 'population at risk'*

In addition to the methods of comparing point patterns suggested above, a number of other techniques have been proposed. One very specific problem is that of comparing a pattern of disease incidence with the distribution of an underlying population at risk. A well-known technique for this is the so-called *Geographical Analysis Machine* (Openshaw et al., 1987). Essentially, this operates by generating a regular grid of points over some study area, and comparing levels of rare disease incidence in circular regions centred on these points with a reference Poisson distribution based on the population at risk in these circles. Whenever an excessively high count of cases occurs (usually defined in terms of a local significance test), the offending circle is tagged on a map. In this way, tentative local clusters can be identified.

Although originally presented in a formal inferential framework, the method is now more widely regarded as an exploratory technique – see for example Fotheringham and Zhan (1996) – mainly due to the problems of interpreting a large number of non-independent significance tests.[5] Attempts have also been made to provide a more rigorous formal adaptation of the approach (Besag and Newell, 1991).

Another methodology is to use a case-control study, so that a set of members of a population at risk without the rare disease are compared with a set of members having the disease (Diggle et al., 1990). Here, an attempt is made to identify any differences between the spatial distribution of the two samples. Comparisons of this kind could be made using the methods set out in Section 6.6.1.

6.7 Conclusions

In this chapter, a number of possible probabilistic models for geographical point processes have been discussed. An obvious way of grouping these models is in terms of first- or second-order characteristics. The first type of model assumes independence of individual events, but non-uniform intensity. The second assumes uniform intensity but non-independence. Deciding which kind of model to apply depends strongly on the context of the study, since for a single data set it is not possible to distinguish between the two types of process on the basis of that data alone. However, if one is simply testing for 'clustering', often first- and second-order-based approaches will tend to give similar conclusions.

The situation is more complex if one is comparing distributions. As the burglary example has shown, two distributions with different first-order properties can have similar second-order properties. Again, it is difficult to tell whether the two burglary groups (the drillers and non-drillers) are both taken from the same second-order process, or from two distinct first-order processes by just looking at the data. As before, one needs to consider the model in the context of the study to decide which explanation is most plausible.

Finally, it is worth noting that, in all cases, visualization and spatial data exploration play an important role. Whether one is estimating K functions, estimating kernel density surfaces, or just looking at the 'raw' point pattern, the output is essentially graphical. Indeed, one way of viewing point pattern analysis is as a means of visually enhancing raw data points so that spatial patterns may become more apparent.

Notes

1. There are good reasons not to: some spatial patterns can show through a wide range of different areal units, particularly if the scale of the pattern is much larger than the scale of the individual zones.
2. Note that, traditionally, point pattern analysis does not include the analysis of *continuous* attribute variables – this is covered in other subject areas, such as *kriging* or *geographically weighted regression*.
3. That is, a set of curves produced from a random sample having a 95% probability of containing the true curve.
4. Drilling inferior lock barrels with a battery drill often causes the lock to disintegrate, so entry can be gained to a house without forcing any doors or windows.
5. See for example Chapter 8 for a discussion of this.

7 Spatial Regression and Geostatistical Models

7.1 Introduction

One of the core techniques of the so-called quantitative revolution in geography was that of *linear regression*. Indeed, its use pre-dates the revolution – in physical applications by some three decades (Rose, 1936), and in human geography by at least one decade (McCarty, 1956). However, this was often rightly criticized for being an essentially aspatial method, and therefore inadequate for modelling geographical processes. Gould (1970) states:

> there may be occasions when conventional inferential statistics can serve as research aids. These cases will probably be simple, non-spatial situations in which some educated guesses are capable of being investigated in inferential terms. Such occasions are likely to be rare.

Why was conventional regression seen as an inadequate tool for geographers? Some further insight can be gained by considering Hepple (1974):

> A glance at any set of maps of geographical phenomena will give a strong disposition against a classical urn-model generation of the pattern (and on which classical statistical inference is predicated).

Some of the earliest attempts to provide more genuinely geographical approaches to data analysis were responses to these and similar criticisms. Examples include Ord (1975), Cliff and Ord (1973) and Hordijk (1974). These approaches may generally be described as *spatial regression models*. Although these are now relatively old ideas, they may be regarded as a turning point towards the modern view of spatial data analysis, and are therefore worthy of discussion here. Techniques of this sort have been applied to a wide range of study areas – for example, the spatial distribution of radon gas (Vincent and Gatrell, 1991), hydrology (Bras and Rodriguez-Iturbe, 1985), epidemiology (Cook and Pocock, 1983), geographical patterns in unemployment (Molho, 1995) and international trade (Aten, 1997). The range of dates for the preceding list of studies suggests that this kind of model continues to influence a broad span of disciplines.

To understand the shortcomings of ordinary linear regression and the theoretical contribution made by spatial regression models, consider the following practical example. Figure 7.1 shows the percentage of households that are owner occupied in wards in the county of Tyne and Wear, taken from the 1991 UK Census of Population. From this, it is clear that owner occupation tends to have higher rates on the periphery of the county, with the lowest rates being observed in the central area.

From the same source, a map of ward-based male unemployment rates in the same area is given in Figure 7.2. Note a general reversal of the pattern here – the *highest* rates of unemployment are in the centre. Suppose we went to investigate the linkage between unemployment and owner occupation. In particular, we would like to investigate whether male unemployment is a good predictor of owner occupation.

After some investigation of variable transforms, a scatterplot of the square root of male unemployment against owner occupation (Figure 7.3) suggests that, at least within the ranges in which the two variables commonly occur, a linear relationship exists between the two quantities. That is, after a square root transformation of the unemployment variable, one could model the relationship between the two quantities using linear regression. Therefore, a model of the form

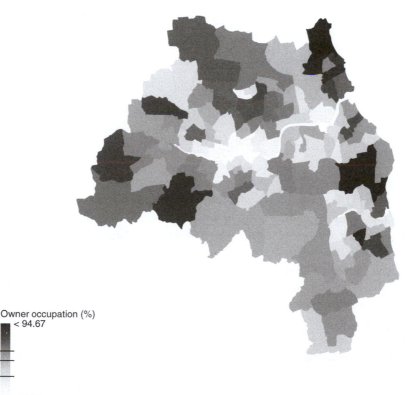

Owner occupation (%)
< 94.67

< 6.32

Figure 7.1 **Owner occupation in Tyne and Wear**

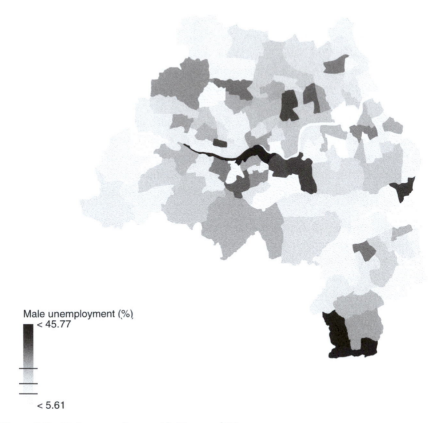

Male unemployment (%)
< 45.77

< 5.61

Figure 7.2 **Male unemployment in Tyne and Wear**

$$\text{Owner occupation} \sim \text{N}(\beta_0 + \beta_1 \sqrt{\text{male unemployment}}, \sigma) \text{ (independently)}$$
$$(7.1)$$

is proposed. The values of β_0 and β_1 can be estimated using ordinary least squares regression. This gives the results shown in Table 7.1. Note that although it may seem strange that $\beta_0 > 100$ for the range of the explanatory variables encountered in the data set, the predicted levels of owner occupation lie within zero and 100% – the linear approximation seems reasonable *within the range of the data*. After fitting this model, the residuals are mapped. Residuals here are defined as the true value of owner occupation minus the fitted value. The resultant map is shown in Figure 7.4.

So what is wrong with proceeding in this way? The main problem is made apparent by the residual map. Note that the values of the residuals do not appear to vary randomly over space – high-valued residuals (darker areas on the map) seem to occur near to each other, as do low-valued residuals, and there are some quite sudden changes from the highest to the lowest values. This is at variance with the assumption in (7.1) that each observation is independent of the others, and in

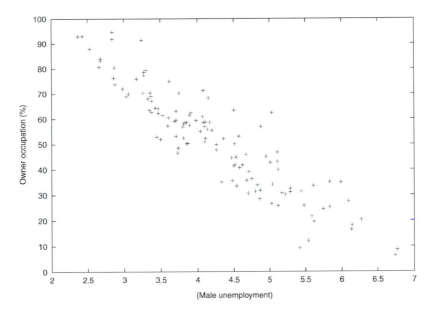

Figure 7.3   **The square root of male unemployment vs. owner occupation in Tyne and Wear**

Table 7.1   **Results of standard regression model for owner occupation data**

Parameter	β_0	β_1
Estimate	131.60	−18.80
Standard error	3.34	0.768

concordance with the comments of Gould (1970) and Hepple (1974) cited earlier in this section. If the independence assumption was true the error terms in the regression model could be modelled as uncorrelated random Gaussian variables, but here it seems that nearby terms tend to take the same sign. If one cannot assume independence in the error terms, then the least-squares calibration of β_0 and β_1 is no longer justified. The remainder of this chapter is concerned with the methods of addressing this. These fall into three broad categories, as shown in Table 7.2. The term *usual* data type is used here as it is possible to apply point-based techniques to area data, for example by replacing each zone by its centroid, and also to apply area-based techniques to point data, for example by computing Voronoi tessellations to a set of points. Note also that non-independent error terms may be modelled either by considering the correlation between error terms, or by modelling a spatial trend in the error terms, rather than assuming the expected value of the error is everywhere zero.

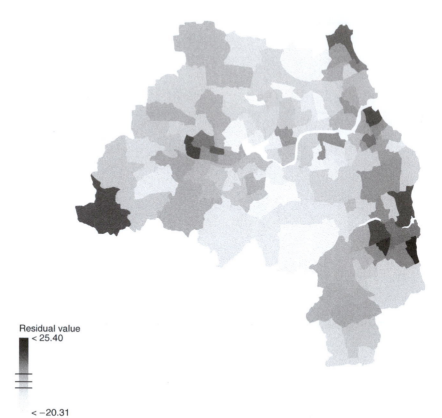

Residual value
< 25.40

< −20.31

Figure 7.4   **Map of residuals for regression model (7.1)**

Table 7.2   **Techniques considered in this chapter**

Method type	Usual data type	Model form
Autoregressive models	Area based	Correlated error terms
Kriging	Point data	Correlated error terms
Smoothing	Point data	Spatial trend

7.2   Autoregressive models

In more general terms, the model (7.1) can be stated in terms of n observations of a dependent variable, contained in the vector $\mathbf{y}$, and n observations of m dependent variables contained in the n by m matrix $\mathbf{X}$. In this case, the ordinary regression model can be expressed in the form

$$\mathbf{y} = \mathbf{X}\boldsymbol{\beta} + \epsilon \qquad\qquad (7.2)$$

where $\boldsymbol{\beta}$ is a vector of regression coefficients, and ϵ is a vector of random errors. Here, the elements of ϵ are independently distributed about a mean of zero. Thus, for the ordinary model one may state

$$\epsilon \sim \text{MVN}(\mathbf{0}, \sigma^2 \mathbf{I}) \qquad (7.3)$$

where $\mathbf{I}$ is the identity matrix and $\mathbf{0}$ is the zero vector, and $\text{MVN}(\mathbf{M}, \mathbf{S})$ denotes a multivariate normal distribution with mean vector $\mathbf{M}$ and variance–covariance matrix $\mathbf{S}$. The problem with this model is the assumption that the variance of ϵ is $\sigma^2 \mathbf{I}$. A more general approach would allow for a broader class of variance–covariance matrices, which would allow for non-independence in the residuals. Most general in this sense would be to assume a model of the form

$$\epsilon \sim \text{MVN}(\mathbf{0}, \mathbf{C}) \qquad (7.4)$$

where $\mathbf{C}$ is any valid variance–covariance matrix. This would resolve the general problem of the assumption of independence, but creates two new problems. First, Figure 7.4 implies that there is a spatial pattern in the dependence of residuals. The general covariance matrix $\mathbf{C}$ does not specifically reflect this. Secondly, although it is easy to calibrate $\boldsymbol{\beta}$ if $\mathbf{C}$ is known, $\hat{\boldsymbol{\beta}} = (\mathbf{X}'\mathbf{C}\mathbf{X})^{-1}\mathbf{X}'\mathbf{C}\mathbf{y}$ (Mardia et al., 1979), normally this is not the case and so both $\mathbf{C}$ and $\boldsymbol{\beta}$ must be calibrated from the sample data.

7.2.1 Spatially autoregressive models

The problem of reflecting *spatial* dependency in the model will be considered first. One way to think of this is to postulate that for zone-based data, the random components of the model in adjacent zones are likely to be linked. For example, if considering house owner occupation rates, it is quite likely that the rate in a zone will mimic to some extent the rates of surrounding zones – especially if the zones are fairly geographically compact – say, one or two neighbourhood blocks. One way to account for this in a regression model would be to use, as an extra explanatory variable, the mean of the dependent variable for adjacent zones. Thus, in the house ownership example, for each zone we would compute the mean of the rates for adjacent zones and use this as an explanatory variable. Returning to the algebraic notation, if $\mathbf{y}$ is the dependent variable, we need to transform this into a vector of adjacent means. This is a linear operation, and can be written as $\mathbf{Wy}$, where $w_{ij} = 0$ if zones i and j are not adjacent, and $w_{ij} = 1/a_i$ if i and j are adjacent, and a_i is the number of zones adjacent to zone i. In fact $\mathbf{W}$ is just the adjacency matrix for the zones, with rows rescaled to sum to one. Zones are not assumed to be adjacent to themselves, so $w_{ii} = 0$ for all i. If the adjacent-mean variable is added to the regression model, the modified model becomes

$$y = \mathbf{X}\boldsymbol{\beta} + \rho\mathbf{W}\mathbf{y} + \epsilon \qquad (7.5)$$

where ρ is the regression coefficient for the adjacent-mean variable. For obvious reasons, this is often called an *autoregressive* model. Subtracting $\rho\mathbf{W}\mathbf{y}$ from both sides of (7.5) gives

$$\mathbf{y} - \rho\mathbf{W}\mathbf{y} = \mathbf{X}\boldsymbol{\beta} + \epsilon \qquad (7.6)$$

Factoring the left hand side gives

$$(\mathbf{I} - \rho\mathbf{W})\mathbf{y} = \mathbf{X}\boldsymbol{\beta} + \epsilon \qquad (7.7)$$

and, assuming $(\mathbf{I} - \rho\mathbf{W})$ is invertible, pre-multiplying both sides by this expression gives

$$\mathbf{y} = (\mathbf{I} - \rho\mathbf{W})^{-1}\mathbf{X}\boldsymbol{\beta} + (\mathbf{I} - \rho\mathbf{W})^{-1}\epsilon \qquad (7.8)$$

Thus, after transforming the $\mathbf{X}$ matrix, the model is of a similar form to (7.2), but with an error term that is a linear transformation of the original, independent vector ϵ. In fact the new error term has variance–covariance matrix

$$\mathbf{C} = \sigma^2[(\mathbf{I} - \rho\mathbf{W})^{-1}]'(\mathbf{I} - \rho\mathbf{W})^{-1} \qquad (7.9)$$

If $\mathbf{C}$ is defined as a function of ρ as above, then it will reflect a spatial structure in error dependencies. Thus, once ρ is known, $\mathbf{C}$ can be computed and $\boldsymbol{\beta}$ can be estimated. The problem is now one of estimating ρ from the sample data.

This problem can be divided into two parts – first finding a valid range for ρ and secondly, finding an *optimum* ρ in terms of a maximum likelihood criterion. The first part is important because the variance–covariance matrix $\mathbf{C}$ must be valid – not all n by n matrices of real numbers meet this condition. The validity condition may be stated as below:

$$\sum_i \sum_j \alpha_i \alpha_j c_{ij} \geq 0 \qquad (7.10)$$

where the α_i and α_j are arbitrary real numbers. This is basically a condition that all linear combinations of the elements of $\mathbf{y}$ have a positive (or zero) variance. It can be shown (see for example Griffith, 1988) that for the autoregressive model this condition is satisfied if and only if $|\rho| \leq 1$.

The second part of the problem can be approached by considering the expression for the likelihood of **y** given values of $\boldsymbol{\beta}$ and ρ. This is monotonically equivalent to

$$\sum_i \left(1 - \rho \sum_j w_{ij} y_j - \sum_j \beta_j x_{ij} \right)^2 \tag{7.11}$$

Finding $\boldsymbol{\beta}$ and ρ to minimize this expression is equivalent to finding maximum likelihood estimates. The optimization process can be broken down into two parts:

1 Find the optimum ρ.
2 Estimate $\boldsymbol{\beta}$ by substituting the *estimate* of ρ into the standard equation.

A number of calculation 'tricks' may be applied to make the first stage more computationally efficient; see Ord (1975) for a discussion of this. An alternative, but sometimes much less reliable, approach is to estimate $\boldsymbol{\beta}$ and ρ using least squares, although this is not the true maximum likelihood in this case. The consequence of this is that the least-squares estimates are often biased.

Applying the maximum likelihood methods to the owner occupation data gives the results of Table 7.3. Note that there are a few changes from the results for the standard regression. First, the value for β_0 is a little lower and the value for β_1 is a little higher, suggesting a smaller negative slope on the linear regression. The estimated value for ρ suggests that some small degree of autocorrelation occurs, implying that neighbouring rates of owner occupation have some effect on the observed rate for a given zone. Finally, note that the standard errors for the estimates are a little larger than in the simple model. This is quite typical behaviour when extra parameters are introduced into models. In this case, the single parameter ρ has been introduced.

Table 7.3 **Results of autoregressive model for owner occupation data**

Parameter	β_0	β_1	ρ
Estimate	123.31	−18.11	0.10
Standard error	6.85	0.96	0.07

7.2.2 Spatial moving average models

A variation on the autoregressive model is the *moving average* model. In this model, it is not the dependent variable that is considered as autoregressive, but the *error term*. The regression model is thus

$$\mathbf{y} = \mathbf{X}\boldsymbol{\beta} + \mathbf{u}$$
$$\text{where} \quad \mathbf{u} = \rho\mathbf{W}\mathbf{u} + \epsilon \qquad (7.12)$$

The inspiration for this model comes from time series analysis, where both autoregressive and moving average models may be used to model the behaviour of a variable observed at regular time intervals – see for example Kendall and Ord (1973). In many practical cases it is difficult to see whether this or an autoregressive approach gives a more accurate reflection of the spatial process under examination. One theoretical difference between the two models is that whereas in the autoregressive model it is typical that *all* error terms are correlated (albeit in a manner which reduces as zones become further apart), in the moving average model error terms are only correlated to their immediate neighbours, as specified in **W**. In more formal terms, for moving average processes, **C** is a sparse matrix if **W** is based on contiguities between zones.

Calibrating moving average models is somewhat harder than calibrating autoregressive models. The simplest approach is to note that Equation (7.12) may be rearranged (in a similar manner to Equations (7.5)–(7.8)) to give

$$(\mathbf{I} - \rho\mathbf{W})\mathbf{y} = (\mathbf{I} - \rho\mathbf{W})\mathbf{X}\boldsymbol{\beta} + \epsilon \qquad (7.13)$$

Thus, if ρ were known, $\boldsymbol{\beta}$ could be estimated by an ordinary least-squares regression of a linear transform of the **y** variables against the same transform of the **X** variables. Of course, ρ is unknown. However, an estimate of ρ could be obtained by calibrating an autoregressive model on the residuals of the ordinary least-squares model, that is finding the value of ρ minimizing $\epsilon'\epsilon$ in Equation (7.12). Once this is done, and a new $\boldsymbol{\beta}$ estimate is obtained, one can re-estimate the residuals, and then iterate the entire process. Repeating this can be shown to lead to a converging sequence of $\boldsymbol{\beta}$ and ρ estimates. Thus, the entire procedure may be set out as below:

1   Obtain an initial estimate of $\boldsymbol{\beta}$ using ordinary least squares:

$$\hat{\boldsymbol{\beta}} = (\mathbf{X}'\mathbf{X})^{-1}\mathbf{X}'\mathbf{y}$$

2   Use the result from step 1 to estimate a set of residuals. Obtain an initial estimate of ρ from these, using the autoregressive model calibration technique. Call this $\hat{\rho}$.

3   Use the current estimate of ρ to update the $\boldsymbol{\beta}$ estimate, given $(\mathbf{I} - \hat{\rho}\mathbf{W})\mathbf{y} = (\mathbf{I} - \hat{\rho}\mathbf{W})\mathbf{X} + \epsilon$.

4   Use the updated $\boldsymbol{\beta}$ estimate to obtain an updated set of residual estimates, and then an updated ρ estimate. Return to step 3 and iterate until $\hat{\rho}$ and $\hat{\boldsymbol{\beta}}$ converge.

Applying this procedure to the owner occupation data gives the results set out in Table 7.4. Here, the value of ρ is much larger than before, suggesting a strong degree of autocorrelation in the error term. Also, although the standard errors are still larger than those in the standard model (Table 7.1) they are considerably smaller than those seen in Table 7.3. It seems there is a stronger case for a spatial effect in the error term than there is for an autoregressive effect. Typically, this suggests that deviations from a global model take place at a very local scale – a suggestion which is also supported by the residual map.

7.3 Kriging

7.3.1 The statistical technique

The above section deals with approaches to spatial modelling for zone-based data, where *proximity* is represented by **W**, the adjacency matrix. However, with point-based data, the concept of adjacency is not so implicitly defined. A more natural approach here might be to consider **D**, a matrix of distances between the points, so that d_{ij} is the distance between points i and j. This methodology plays a core role in the field of *geostatistics*. Suppose there is a set of n points with a distance matrix as defined above, also with a vector of attached dependent variables **y** and a set of attached independent variables, **X**. As before, it is intended to calibrate a regression model $\mathbf{y} = \boldsymbol{\beta}\mathbf{X} + \epsilon$, where the vector ϵ has distribution MVN(**0**, **C**), and it is required that the elements of **C**, the variance–covariance matrix of the error terms, reflect the spatial structure of the data.

A possible way of achieving this is to let c_{ij} be some function of d_{ij}, say $c_{ij} = f(d_{ij})$. This would mean that the covariance of a pair of error terms would depend on the distance between them. Typically, one would expect f to be a decreasing function, so that the linkage between error terms decreased as the distance between them increased. For this, the positive definiteness condition on the matrix **C** discussed earlier becomes important once again. Recall that this required

$$\sum_i \sum_j \alpha_i \alpha_j c_{ij} \geq 0$$

Table 7.4 **Results of moving average model for owner occupation data**

Parameter	β_0	β_1	ρ
Estimate	137.2	−20.08	0.65
Standard error	3.94	0.79	0.10

for all α_i and α_j. In the current situation, this requires that

$$\sum_i \sum_j \alpha_i \alpha_j f(d_{ij}) \geq 0$$

Unfortunately, this means that one cannot choose an arbitrary function f as most do not satisfy the condition. There are, however, some functional forms which do meet this requirement. A number of possible candidates include the exponential form

$$c_{ij} = \sigma^2 \exp(-d_{ij}/h) \qquad (7.14)$$

or the power form

$$c_{ij} = \sigma^2 \exp(-d_{ij}^2/h)^2 \qquad (7.15)$$

or the spherical form

$$c_{ij} = \begin{cases} \sigma^2 \left(1 - \dfrac{3d_{ij}}{h} + \dfrac{d_{ij}^3}{h^3}\right) & \text{if } d_{ij} < h \\ 0 & \text{if } d_{ij} \geq h \end{cases} \qquad (7.16)$$

In each case, h acts in a similar manner to the bandwidth in kernel density estimation (see Chapters 4 and 6), and determines the distance around a given observation over which other observations are likely to be dependent: the parameter h is usually expressed in the same distance units as the distances between the points, d_{ij}. In each of the above equations there are two unknown parameters, σ and h. In simple terms σ describes *how variable* the error term is, and h describes *how spatially influential* it is. Note also that by dividing any of (7.14)–(7.16) by σ^2 an expression for the *correlation* between ϵ_i and ϵ_j is obtained. For example, for Equation (7.14) we have

$$\text{corr}(\epsilon_i, \epsilon_j) = \exp(-d_{ij}^2/h^2) \qquad (7.17)$$

The three different functional forms for the relationship between correlation and distance are shown in Figure 7.5. For each curve, $h = 1$. Note that although in general wider bandwidths imply increasing radii of non-trivial correlation, the same value of bandwidth for different functions does not imply exactly the same radius of influence for both functions. Note that in all of these functions, as d_{ij}

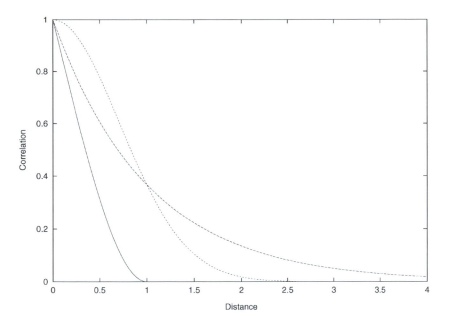

Figure 7.5 **Graphs of correlation functions: the light dotted line is for model (7.14) the heavier dotted line is for (7.15) and the solid line is for (7.16); in each case $h = 1$**

approaches zero, the covariance approaches σ^2. However, in some cases this may be inappropriate – there will be some 'extra' variability at the point of observation perhaps due to measurement or sampling error. Here the limiting covariance as d_{ij} approaches zero is some quantity τ^2, say, which is not equal to σ^2. In this case, the covariance function undergoes a jump of size $\sigma^2 - \tau^2$ at the point $d_{ij} = 0$. This is sometimes referred to as the *nugget effect*. For example, Equation (7.14) can be extended to the form below:

$$c_{ij} = \begin{cases} \tau^2 \exp(-d_{i}j/h) & \text{if } d_{ij} > 0 \\ \sigma^2 & \text{if } d_{ij} = 0 \end{cases} \qquad (7.18)$$

Graphically, this relationship would appear much the same as the curve for Equation (7.14) shown in Figure 7.5, but with a single discontinuity at $d_{ij} = 0$.

The technique of *kriging* (Krige, 1966) involves finding estimates for the regression coefficients β as well as for σ, τ and h. This problem is not as simple as for ordinary least-squares regression, since one needs to know σ, τ and h in order to estimate β, but also one needs to know β in order to estimate σ, τ and h! A compromise is to start with a *guess* at β, use this to estimate σ, τ and h and then to re-estimate β in the light of this. A plausible initial guess for β would be the ordinary least-squares estimate.

From this starting point, one can obtain a set of residuals from the observed y_i

variables. Call these e_i. These may be used as estimates for the true residuals ϵ_i. From these, we need to estimate the parameters σ, τ and h, by considering the spatial relationship between the residual pairs e_i and e_j for points i and j, and the distance between the points d_{ij}. A useful way of measuring the difference between e_i and e_j is to consider the squared difference $(e_i - e_j)^2$. Consider the expected value of this expression, $E[(e_i - e_j)^2]$. Expanding, we have

$$E[(e_i - e_j)^2] = E(e_i^2) - 2E(e_i e_j) + E(e_j^2) \qquad (7.19)$$

but, given that e_i and e_j have expected values of zero, we can state that $E(e_i^2) = E(e_j^2) = \sigma^2$ and $E(e_i e_j) = \text{cov}(e_i, e_j) = c_{ij}$, so that

$$E[(e_i - e_j)^2] = 2\sigma^2 - 2c_{ij} \qquad (7.20)$$

Thus, the expected value of $(e_i - e_j)^2$ is very simply related to the covariance function, and can be expressed as a function of d_{ij}:

$$E[(e_i - e_j)^2] = 2\sigma^2 - 2f(d_{ij}) = g(d_{ij}) \qquad (7.21)$$

and $g(d_i j)$ can therefore be expressed as a function with parameters σ, τ and say, h. For example, if the covariance model (7.14) is adopted, then $g(d_{ij}) = \sigma^2[2 - 2\exp(-d_i j/h)]$. Adding a nugget effect, we have

$$g(d_{ij}) = \begin{cases} (\sigma^2 - \tau^2)2 - [2\exp(-d_i j/h)] + \tau^2 & \text{if } d_{ij} > 0 \\ \sigma^2 & \text{if } d_{ij} = 0 \end{cases} \qquad (7.22)$$

Thus, by fitting a non-linear curve to $(e_i - e_j)^2$ as a function of d_{ij}, it is possible to estimate h, σ and τ. Sometimes the function g is divided by 2, since this results in the right hand side of the above equation losing an 'untidy' factor of 2. Noting that $E[(e_i - e_j)^2] = \text{var}(e_i - e_j)$, this halved version of g is sometimes called the *semi-variance*, and a graph of g is called a *semi-variogram*. This is a particularly useful technique if the nugget effect is present, since σ^2 can be more readily estimated as the asymptotic value of the semi-variogram as d_{ij} tends to infinity.

Calibrating the semi-variogram is itself a complex task. Even in a situation where the ϵ_i are independent, the e_i are not (Dobson, 1990), and so one cannot expect the $(e_i - e_j)^2$ to be. However, in general kriging practice this problem is ignored, and h and σ are usually estimated using non-linear least-squares approaches – that is, finding h and σ to minimize

$$\sum_{i,j}[(e_i - e_j)^2 - g(d_{ij}|h, \sigma, \tau)]^2 \qquad (7.23)$$

where $g(d_{ij}|h, \sigma, \tau)$ denotes the dependence of g on the three parameters. Having found reasonable approximations for h and σ, one can then obtain estimates for the c_{ij}, and proceed to recalibrate the model for β. If, as before, we denote the entire covariance matrix by C, we then have

$$\hat{\beta} = (X'CX)^{-1}X'Cy \qquad (7.24)$$

as discussed in earlier sections when estimating β in situations where there is correlation between error terms. This rather lengthy procedure can be summarized as follows:

1   Estimate β using ordinary least squares.
2   Estimate residuals from this β estimate.
3   Calibrate the semi-variogram from the residuals.
4   Use the calibration of the semi-variogram to estimate C.
5   Re-estimate β using $(X'CX)^{-1}X'Cy$.

7.3.2   A worked example

The technique of kriging is best illustrated with a worked example. Here the data described earlier will be used. Although this is zone rather than point data, one can attribute the data to ward centroids. In this way it is possible to compare the results of this analysis with those of previous sections in this chapter. As with previous analyses, the explanatory variable here will be the *square root* of unemployment, as this appears to have a linear relationship with owner occupation.

To recap, the ordinary least-squares estimates for the elements of β were $\beta_0 = 131.6$ and $\beta_1 = -18.8$. From this, residuals were computed, as shown earlier in Figure 7.1. This supplies us with a set of residuals which may be used to estimate h and σ. In this case, there are 120 wards, and therefore 120 centroids. Thus, there are 7140 d_{ij} values to consider, together with the same number of values of $(e_i - e_j)^2$. For a non-linear regression problem this may be computationally intense. Therefore, another common computational shortcut used in kriging is to 'bin' the d_{ij} categories into a series of distance intervals, and consider the average $(e_i - e_j)^2$ values for each category as a function of the central distance values for that category. In some cases, the bin categories might overlap (Bailey and Gatrell, 1995). Here bins of width 2 km centred on the d_{ij} values 1 km, 2 km, ..., 16 km were used, and a semi-variogram curve of the form (7.22) was fitted to the observed average $(e_i - e_j)^2$ values. The results of this exercise are shown in Figure 7.6. Here, the fitted parameter values are $h = 5.6$ km, $\sigma^2 = 160.8$ and $\tau^2 = 46.75$. From

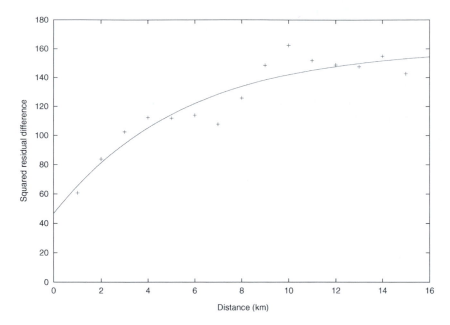

Figure 7.6   Results of fitting a theoretical semi-variogram of the form in model (7.22)

these, we can construct the **C** matrix, and re-estimate β_0 and β_1. The re-estimates are shown in Table 7.5. This gives a very similar value of β_0 and β_1 to the spatial moving average model.

7.3.3   Trend surfaces from kriging residuals

One very useful product of the kriging approach to regression is that it is possible to predict the value of residuals at points other than those at which the observations are taken. Suppose a new point is added to the data set, at a location (u, v). Without any other information, the mean value of the error term expected at this point is zero. However, if there is spatial autocorrelation amongst the error terms, and the values of some errors *near* to the new point are known, then this situation changes. For example, if many nearby error terms are positive, one might reasonably expect the error at (u, v) to be positive also. Essentially, we are now considering the *conditional* probability distribution of the error at (u, v) *given* the errors at the

Table 7.5   **Results of kriging model for owner occupation data**

Parameter	β_0	β_1
Estimate	138.2	−20.06

other locations. Using this distribution, it is possible to predict the likely error term at (u, v) and therefore apply a correction to the ordinary least-squares predictor of y at (u, v) given the predictor variables.

In practice we do not know the error terms at each point, but we do know the residuals from fitting the regression model after kriging. The latter may be used as an approximation for the former. Next we must consider how one can predict the error at (u, v) given a vector of n errors $\{\epsilon_i\}$, or residuals $\{e_i\}$. One way of addressing this is to consider the *expected mean square error* in predicting the error term. If we denote the true value of the error term at (u, v) by $\epsilon(u, v)$, and the estimated value of this by $\hat{\epsilon}(u, v)$, then we are attempting to minimize

$$E\{[\epsilon(u, v) - \hat{\epsilon}(u, v)]^2\} \tag{7.25}$$

In particular, suppose we consider estimates that are linear combinations of the e_i, so that

$$\hat{\epsilon}(u, v) = \sum_i \gamma_i e_i \tag{7.26}$$

Then, it can be shown (Bailey and Gatrell, 1995), that if γ is the vector of the γ_i in the above expression then (7.25) is minimized when

$$\gamma = \mathbf{C}^{-1}\mathbf{c}(u, v) \tag{7.27}$$

where $\mathbf{C}$ is the n by n covariance matrix between the errors at the n observed points, and $\mathbf{c}(u, v)$ is the n by 1 vector of covariances between the error at (u, v) and the error at the n observed points. Recall that in the kriging procedure we calibrated a functional relationship between the covariance of a pair of error terms and the distance between the points at which the error terms occurred. This not only allows an estimate of $\mathbf{C}$ to be made as before, but also allows an estimate of $\mathbf{c}(u, v)$ to be made, based on the distances between (u, v) and the n observed point locations. Thus, if $\mathbf{e}$ is a vector of the residuals $\{e_i\}$, then we have

$$\hat{\epsilon}(u, v) = (\mathbf{C}^{-1}\mathbf{c}(u, v))'\mathbf{e} \tag{7.28}$$

Thus, it is possible to predict the error term (or at least compute its conditional mean value) at any point (u, v) in the study area. Thus, it is possible to consider spatial trends in the error term, and by mapping $\hat{\epsilon}(u, v)$ identify regions where the kriging regression model is likely to over- or underpredict. For the home ownership application discussed above, a map of $\hat{\epsilon}(u, v)$ is shown in Figure 7.7. One notable

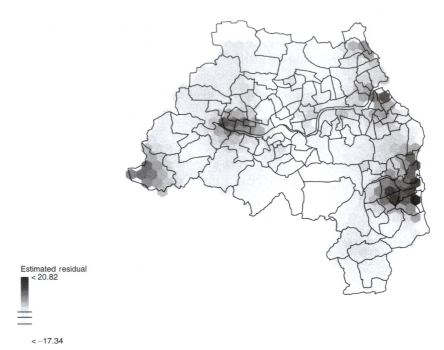

Figure 7.7 **Map of $\hat{\epsilon}(u,\ v)$ for owner occupation example in Section 7.3.2**

feature here is a region of positive predicted residuals in the south-eastern part of the map. This region roughly corresponds to the city of Sunderland. A number of other prominent features can also be identified, for example around the coastal community of South Shields, just to the south of the mouth of the River Tyne.

7.4 Semi-parametric smoothing approaches

A final method which models spatial deviations from a simple regression model is the *semi-parametric* model (Hastie and Tibshirani, 1990). This differs from the other models in that it does not assume that the error terms are correlated. Here the model takes the form

$$y_i = f(u_i,\ v_i) + \sum_j \beta_j x_{ij} + \epsilon_i \tag{7.29}$$

The model is termed semi-parametric since one makes no assumptions about the functional form of f. In fact f is used to model spatial deviations from the simple, non-spatial regression model. A plot of $f(u,\ v)$ as a surface in the study area serves a similar purpose to that of Figure 7.6, in that it indicates spatial trends in over- or underprediction of the simple regression model.

To calibrate a model of this kind, a very different approach is required. First, suppose that the error terms $\{\epsilon_i\}$ were known. One could estimate f by applying a spatial smoothing to these quantities. This is achieved at point (u, v) by computing a weighted average of the ϵ_i, where the greatest weight is assigned to observations closest to (u, v). For example, the Gaussian weighting scheme could be defined by

$$w_i(u, v) = k \exp[-d_i^2(u, v)/h^2] \tag{7.30}$$

where $w_i(u, v)$ is the weight applied to the ith residual to estimate $f(u, v)$, and $d_i(u, v)$ is the distance between the ith observation and the point (u, v), h is a bandwidth controlling the degree of smoothing as discussed in Chapters 2, 4, 5 and 6, and k is a normalizing constant to ensure that $\sum_i w_i(u, v) = 1$. Using these weights, we can estimate f at (u, v) using

$$\hat{f}(u, v) = \sum_i w_i(u, v)\epsilon_i \tag{7.31}$$

It is worth noting the similarity between this approach and the geographically weighted regression (GWR) technique described in Chapter 5 (Brunsdon et al., 1996). As with kriging, we do not have observed values of ϵ_i. However, if $\boldsymbol{\beta}$ were known we could use the residual terms $\{e_i\}$ as estimates, giving a revised estimate

$$\hat{f}(u, v) = \sum_i w_i(u, v)e_i \tag{7.32}$$

The problem here is that we do not know $\boldsymbol{\beta}$. However, rearranging (7.29) and replacing f by its estimate $\hat{f}$ we have

$$y_i - \hat{f}(u_i, v_i) = \sum_j \beta_j x_{ij} + \epsilon_i \tag{7.33}$$

Thus, $\boldsymbol{\beta}$ could be estimated by regressing $y_i - \hat{f}(u_i, v_i)$ on the explanatory variables. Obviously, this estimate, say $\hat{\boldsymbol{\beta}}$, could then be used in Equation (7.32). In matrix form, the relationship between $\hat{f}$ and $\hat{\boldsymbol{\beta}}$ can be expressed by the simultaneous equations

$$\hat{\mathbf{f}} = \mathbf{W}(\mathbf{y} - \mathbf{X}\hat{\boldsymbol{\beta}})$$
$$\hat{\boldsymbol{\beta}} = (\mathbf{X}'\mathbf{X})^{-1}\mathbf{X}'(\mathbf{y} - \hat{\mathbf{f}}) \tag{7.34}$$

where $\mathbf{W}$ is the matrix whose ijth element is $w_i(u_j, v_j)$ and $\hat{\mathbf{f}}$ is a column vector whose ith element is $\hat{f}(u_i, v_i)$. After a lengthy sequence of algebraic manipulation it is possible to solve explicitly for $\hat{\boldsymbol{\beta}}$ and $\hat{\mathbf{f}}$:

$$\hat{\boldsymbol{\beta}} = (\mathbf{X}'(\mathbf{I} - \mathbf{W})\mathbf{X})^{-1}\mathbf{X}'(\mathbf{I} - \mathbf{W})\mathbf{y} \tag{7.35}$$

and

$$\hat{\mathbf{f}} = \mathbf{W}(\mathbf{y} - \mathbf{X}\hat{\boldsymbol{\beta}}) \tag{7.36}$$

so that it is possible to calibrate the entire model using explicit formulae. It is also worth noting the similarity between Equation (7.35) for the semi-parametric approach and Equation (7.24) for kriging. The smoothing approach and the kriging approach can give very similar estimates for $\boldsymbol{\beta}$. In fact if $\mathbf{I} - \mathbf{W} = \mathbf{C}$ then the two approaches give identical estimates.

One final issue to be addressed is the choice of h, the smoothing bandwidth. When h is very small, the spatial smoothing applied to the residuals will be extremely 'spikey' and will follow the random variations in the error terms too closely. However, when h is very large then oversmoothing may occur, so that genuine features in $f(u, v)$ may be 'smoothed out' and not detected in $\hat{f}(u, v)$. Also, selection of h is not akin to estimating a model parameter. The parameters one is trying to estimate are the function f and the vector $\boldsymbol{\beta}$. The most appropriate h for one particular data set will not be the same for another – factors such as the density of the observation points as well as the shape of the unobserved function f will all affect the choice of h. One approach to choosing h is to consider the expected mean square error for the prediction of y, as was done when predicting residuals in kriging. In the semi-parametric case, there is no simple algebraic solution giving h which minimizes this quantity. Instead, one has to rely on estimating this quantity from the data, and find h which minimizes this. There are a number of ways of estimating the expected mean square error from the data – one of these is to compute the *generalized cross-validation score* or GCV score (Hastie and Tibshirani, 1990). This is defined to be

$$\text{GCV} = \frac{1}{n}\sum_{i}\left\{\frac{y_i - \hat{y}_i}{1 - \text{tr}(\mathbf{S})/n}\right\}^2 \tag{7.37}$$

where $\hat{y}_i$ is the fitted value for y_i, $\text{tr}(\mathbf{X})$ denotes the sum of the leading diagonal (or *trace*) of the matrix $\mathbf{X}$, and $\mathbf{S}$ is a matrix such that $\hat{\mathbf{y}} = \mathbf{S}\mathbf{y}$, where $\hat{\mathbf{y}}$ is a column vector of predicted y values. For the semi-parametric model seen here, it can be shown that

$$\mathbf{S} = \mathbf{X}(\mathbf{X}'(\mathbf{I} - \mathbf{W})\mathbf{X})^{-1}\mathbf{X}'(\mathbf{I} - \mathbf{W}) + \mathbf{I} - \mathbf{W} \tag{7.38}$$

As h changes, then in turn $\mathbf{S}$ and GCV alter. To choose h, one has to plot its value against the associated GCV estimate, and find the minimum point on this graph. Note that with a large number of observations, computation of $\mathbf{S}$ becomes a problem, since it consists of n^2 elements. Typically, sparse matrix manipulation algorithms are called for here.

As an example, consider the home ownership data once again. In this case, the model fitted takes the form

$$\text{Owner occupation } = \beta_1 \sqrt{\text{unemployment}} + f(u, v) + \epsilon \tag{7.39}$$

Note that there is no β_0 here, since $f(u, v)$ takes an arbitrary form, and so could contain an additive constant such as β_0. For these data, h is plotted against GCV in Figure 7.8. This graph shows that the lowest prediction mean square error occurs when h is around 0.75 km, although the curve is quite flat around this region, and values in the range 0.75–1.0 km will give more or less similar results in terms of

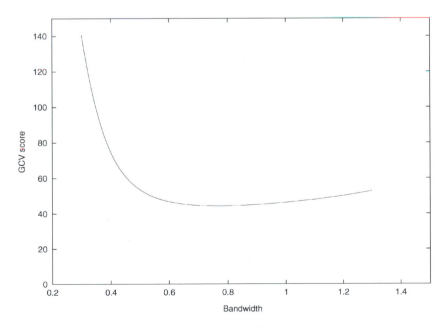

Figure 7.8 **Plot of GCV vs. h for the home ownership model**

predicting the y values. When the optimal value of h is used we obtain an estimate for β_1 of -18.26. The corresponding estimate for $f(u, v)$ is shown in Figure 7.9. It is worth noting the similarity between this map and that of Figure 7.6. Also of interest is the fact that this model could be further extended, by allowing *all* of the coefficients to vary over space. In doing this we would arrive at the GWR approach. Finally, it is possible to extend this model even further by allowing for autocorrelation effects in the model *as well as* spatial variation – and furthermore to allow the value of ρ to vary geographically (Brunsdon et al., 1998a)!

7.5 Conclusions

In this chapter, a number of ways of approaching regression which allow for spatial dependency amongst the observed points are discussed. The first two, the autoregressive and moving average approaches, are essentially zone based. The covariance between error terms is defined in terms of the adjacencies between zones, or the distances between zone centroids. Clearly, this depends very heavily on the choice of areal units. In instances where such information is available, it may be helpful to apply autoregressive or moving average models to the same data aggregated in more than one way. This may give some idea of how sensitive the models are to the areal units used. In fact it is possible to obtain a range of results

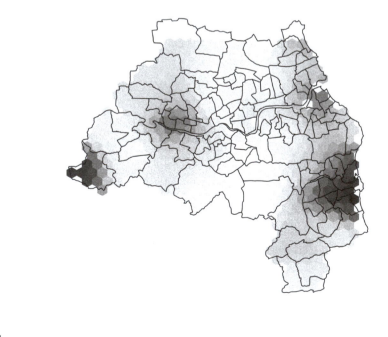

Local term
< 17.32

< −11.43

Figure 7.9 **Map of $\hat{f}(u, v)$ for home ownership model**

with a single data set, by defining adjacency between zones in different ways. This is illustrated in Table 7.6, where the owner occupation moving average model is calibrated first with the adjacency matrix used earlier in the chapter, and secondly with adjacency being defined as a first- or second-order neighbour. This could perhaps be regarded as a new manifestation of the modifiable areal unit problem (Openshaw, 1984), where not only the *scale* and *location* of zones but also the *topology* can influence the results of a statistical analysis. It is worth noting that although it is possible to compute confidence intervals for the model parameters in these models, these do not allow for any uncertainty in the definition of adjacency. In real terms, these confidence intervals are almost certain to be too small.

One way in which kriging differs from the latter techniques is that some attempt is made to calibrate the *extent* of the error term correlation from the observed data. Rather than assuming that dependencies are in terms of immediate zonal neighbours, a correlation function is calibrated based on the distance between points. Once this has been done the regression model is then calibrated using this model of error term correlation. However, as before, this calibration assumes the correlation function to be known exactly, rather than being an estimated curve from residual data. This results in a similar problem to that with autoregressive and moving average models: although confidence intervals can be computed for the regression coefficients, these will most likely be too optimistic since in actuality there is some uncertainty about the correlation function. This is addressed in Diggle et al. (1998), however, using a Bayesian framework (Besag and Green, 1993).

It is interesting to note the similarity between the semi-parametric model and the kriging model. Although these two approaches assume that different processes generated the observed data, they give quite similar predictions. Indeed, by an appropriate choice of correlation function for kriging, and smoothing function for the semi-parametric approach, one can obtain identical predictions. This demonstrates a common problem with spatial regression: it can sometimes be very difficult to infer which underlying process generated the data. However, one positive point is that even if one chooses the wrong process to model, the consequences in terms of prediction error will not be great. In such situations it is perhaps reasonable to adopt the computationally simplest approach, which in this case is perhaps the semi-parametric approach. However, despite the difficulties involved in any of these methods, to the geographer all of these approaches offer an improvement on the simple regression technique, since they all take into account the spatial nature of the process being analysed.

Table 7.6 **Variability of a spatial moving average model as the definition of adjacency is altered**

Adjacency type	β_0	β_1
First-order neighbour only	137.2	−20.08
First- or second-order neighbour	125.2	−18.52

8 Statistical Inference for Spatial Data

8.1 Introduction

The title of this chapter may be daunting to some readers. Statistical inference may appear to be a complex subject – and indeed sometimes it is. However, it is important for quantitative geographers to be aware of the underlying issues for a number of reasons. First, inference plays a key role in any quantitative study. In any study, data are collected in order to *infer* something about an underlying process or situation. Even exploratory studies which do not set out to test predefined hypotheses will require some ideas to be inferred from empirical evidence. Secondly, attitudes to inference are changing within the world of statistics. Approaches which would have been used only very rarely a few decades ago are now commonplace – and there is much more debate about the relevance and utility of some established viewpoints. Clearly, the subject matter of such debate will apply to the work of any quantitative geographer using inferential methods.

Thus, this chapter is intended to introduce the quantitative geographer to a number of issues relating to statistical inference, and to provide a background and introduction to some of the debates currently taking place. Approaches to statistical inference can be categorized in a number of ways. The approach in this chapter is to classify approaches to inference as illustrated in Table 8.1. Although this does not provide an exhaustive coverage of approaches it is felt that it does address all current major issues.

The broadest division is perhaps between *formal* and *informal* approaches. Formal approaches are generally used to test hypotheses that have been suggested by theory before the data have been collected. Such hypotheses are stated in mathematical form. When using a formal approach one attempts to quantify the plausibility that the observed data lend to the hypotheses.

Table 8.1 **Categories of inference addressed in this book**

Inference				
Informal			Formal	
EDA	Visualization	Data mining	Bayesian	Classical (theoretical and experimental)

As one might expect, approaches classed here as 'informal' are less rigorous. With these, the aim is to use graphical techniques, or to represent structure in some other way so that patterns in the data become apparent. On discovering such patterns, the analyst then considers theoretical implications in a more subjective way. As before, prescribed hypotheses could be considered, but informal approaches are also very useful as *generators* of hypotheses. Identifying patterns in the observed data can often suggest new research questions.

Table 8.1 may be used as a map of this chapter. First informal inferential techniques will be considered, and then formal techniques. Within each of these broad divisions, the specific approaches enumerated on the lowest row of Table 8.1 will be considered in turn. Following this, there will be some discussion comparing the two formal approaches covered here, particularly in the context of model choice and model calibration. Some discussion of experimental significance testing then follows. Although this fits into the general classical framework, use is made of Monte Carlo simulation to handle analytically intractable problems. This final part of the chapter is intended to give a flavour of some of the debates within the statistical community mentioned earlier.

8.2 Informal inference

One thing that all of the approaches in this section have in common is their dependence on computing power. For example, visualization makes use of the interactive graphical features found on almost all computers at the time of writing this book. In a recent text, Brunsdon (1995a) argues that the availability of microcomputers puts graphical techniques within the capabilities of most researchers, and advocates their use for the analysis of 1991 UK Census data. This situation is then contrasted with that of the previous census in 1981 when personal computers were available, but few had sufficient graphical capabilities to carry out serious visualization tasks. Similarly, more heavily computational approaches – such as data mining – can take advantage of the speed of current computer hardware.

8.2.1 Exploratory data analysis (EDA) and visualization

The first two approaches here are considered only briefly, since they have an entire chapter devoted to them elsewhere in this book. They are considered together as there is some degree of overlap between them. Certainly, visualizing data through scatterplots and histograms may be thought of as a data exploration method. However, in addition to techniques of this kind, descriptive statistics are also useful for data exploration. For example, the *five-number summary* described in Chapter 4 provides a useful 'at-a-glance' indicator of shape for univariate distributions. In general, careful thought about tabulation is also an important part of exploratory data analysis.

Of particular importance to geographers is the use of cartographic techniques. For any spatial data set, it is important to identify geographical patterns. Typical questions that analysts might wish to ask are as follows:

- Are a set of geographical points clustered?
- Are a set of points dispersed? (That is, are they further apart than a random pattern?)
- Does a set of areal data exhibit autocorrelation?
- Are there any spatial outliers?
- Are there any spatial trends?

The first two questions relate to point patterns. The second phenomenon occurs when some kind of 'repulsion' is present. For example, some animals exhibit territorial behaviour, and will not breed too close to other animals of the same species. In some cases, the first two questions have to be addressed in the presence of a base level of clustering. For example, one might wish to answer questions such as 'Do household burglaries cluster in space *above and beyond* the clustering of observed household locations?' Questions of the last kind can be addressed using *cartograms* (Tobler, 1973a; Dorling, 1991).

The third phenomenon relates to areal or zonal data. Positive autocorrelation occurs if neighbouring zones tend to take on similar data values. For example, average house prices for census wards might tend to follow this pattern. More rarely occurring is negative autocorrelation. This term is used to describe a situation in which data for neighbouring zones tend to be more different than one might expect to occur at random. A rather abstract example of this is to consider the squares on a chessboard as a set of zones, with a zero attached to the white squares and a one attached to the black squares. Note that the point patterns in the first two types of question can be translated to areal data by counting points in zones and converting to rates per unit area. However, care must be taken here, as clustered point patterns can aggregate to either positively or negatively autocorrelated areal patterns – see Figure 8.1. This is an example of the *modifiable areal unit problem* (Openshaw, 1984; Fotheringham and Wong, 1991). This problem can have consequences for visualization as well as the more commonly considered effects in formal analysis.

The fourth type of question is often an important part of analysis. In social science (and occasionally in physical science) outlying observations can occur. These can arise for a number of reasons, which may be divided into two broad categories – those due to recording error and those due to genuinely unusual observations. Unfortunately data analysis alone cannot suggest which cause is most likely for any given example, and one needs to look more carefully at the original data and their collection process. However, exploratory data analysis and visualiza-

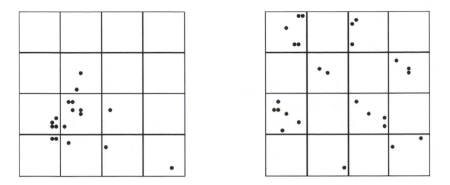

Figure 8.1 **Aggregation effects in clustering. On the left hand side, point data aggregated to the zones shown result in a positively autocorrelated data set. On the right, a negatively correlated aggregate data set results. Both point patterns exhibit clustering**

tion carry out the important task of identifying cases where such consideration may be necessary.

Finally, EDA and visualization may be used to identify trends in data. In a geographical sense, trends could be on an international, national, regional or more local scale. For example, are house prices in the UK generally lower towards the north of the country? Are household burglaries more common (on a per-household basis) in a city centre than in its hinterlands? Trends of this sort can be investigated by mapping data directly, or by using cartographic techniques linked with smoothing techniques, such as GWR (Brunsdon et al., 1996) or probability density estimation (Brunsdon, 1995a).

In summary, using EDA or visualization in any of the above ways can be seen as an inferential process. In each case, one uses an appropriate technique to *infer* an answer to one of the five questions listed above. Clearly, as proposed in Table 8.1 this is an informal approach. In some instances, formal methods may be used to provide more rigorous mechanisms to answer these questions, but these will require a degree of formal model specification before testing can take place. In this case, an initial exploratory examination may suggest model specification strategies. In addition to this, there are other EDA/visualization activities, such as those associated with outlier identification, which are not well replicated by formal methods, but which are nonetheless an important aspect of spatial data analysis.

8.2.2 Data mining

Most visualization and spatial data analysis is applied to a small number of variables at a time. However, in many cases data sets can contain a very large number of variables. If a data set has m variables (or dimensions), then there are $m(m-1)/2$ possible two-way interactions between variables, $m(m-1)(m-2)/6$

three-way interactions, and more generally $m!/k!(m-k)!$ k-way interactions. Although some visualization techniques (such as the scatterplot matrix) do attempt to show two-way interactions, generally other methods are needed to explore the higher-level interactions. One set of approaches, discussed elsewhere in this book, is based on 'projecting' m-dimensional space onto two- or three-dimensional space. Having done this, standard visualization techniques may be applied. Tracing results back to the m-dimensional space will then hopefully identify patterns in the original data.

However, an alternative approach is to use computer algorithms to search for pattern and structure within the data. This approach is sometimes referred to as *data mining*. The data-mining approach encompasses a number of techniques, which make use of a wide variety of underlying ideas. One common characteristic of these approaches is that they require a large degree of computation in comparison with EDA or standard visualization techniques. Another is that despite this computational overhead, data-mining techniques do not provide formal statistical models for the processes under investigation. Two possible techniques will now be considered. The first of these, *cluster analysis*, will be covered in most detail, since it is relatively easily understood, but many of the observations about inference applied to this method are also pertinent to most other forms of data mining. These examples are not intended to be exhaustive, but should provide an idea of the variety of different approaches.

8.2.2.1 Cluster analysis This is perhaps one of the oldest methods which may be classed as data mining. In this approach, a data set is classed into a number of groups of observations, according to some algorithm. As is typical in data-mining applications, there are usually a large number of variables and a large number of observations. Clustering is generally approached in one of two ways: the *hierarchical approach* or the *optimization approach*, although some modern approaches do not fit neatly into either category. Hierarchical methods start off by having each observation in its own group and then repeatedly fuse groups according to some criterion of similarity. For example, in the *single linkage method* the algorithm below is applied:

1 Start off with n groups, with one observation in each group.
2 Find the two groups containing the pair of observations which are closest together and not in the same group. If there are m variables, closeness is generally defined in terms of Euclidean distance in m-dimensional space – but other definitions could be used.
3 Merge the two groups identified in step 2 into one larger group.
4 If there is more than one group remaining, go to step 2.

If this algorithm is carried out to completion, only one group will remain. However, it is often informative to look at the group formation history, and at the sizes of the nearest-neighbour distances associated with the formation of each group. Alternatively, one can modify the stopping rule in step 4 above. For example, the algorithm could terminate when the number of groups remaining fell to some quantity greater than one, or when the nearest-neighbour distance referred to in step 2 exceeded some limit.

The single linkage method is not the only approach to hierarchical clustering. Other approaches can be obtained by modifying step 2 in a number of ways. For example, the nearest-neighbour criterion for choosing a pair of groups to merge could be replaced by an average distance between group members. Clearly, each hierarchical method will produce a set of groups having different properties – and indeed even the same method with different stopping rules will produce a different end result.

The second commonly used method of clustering is the optimization approach. In this approach, the data are partitioned into a number of groups as before, but this is not achieved by the sequential fusion approach adopted in hierarchical clustering. Here, the number of groups is specified in advance, and individual observations are assigned to groups in such a way as to optimize some criterion. This criterion is typically a measure of 'tightness' of the clusters. For example, the *k-means* method attempts to minimize the within-group sum of squares in the grouped data set. This is defined in terms of each group's centroid in *m*-dimensional space, and the squared distance of each group member from this centroid.

This optimization problem is sometimes quite difficult. For example, if one has a data set with 100 observations, and one wishes to attempt a *k*-means clustering with four groups, then there are $4^{100} \approx 1.6 \times 10^{60}$ possible groupings. An exhaustive search of all groupings would not be reasonable, so typically some search heuristic must be adopted. Usually such heuristics work by starting with an initial assignment of observations to groups and then attempting to move observations from one group to another in order to improve the optimization criterion, until no further improvement is possible. Unfortunately, such approaches cannot guarantee that a global optimum will be reached, and the final groupings can depend heavily on the initial groupings. Thus, the output from an optimization clustering algorithm can depend on the initial cluster assignments, the optimization criterion, and the number of clusters specified.

Thus, both hierarchical and optimization-based clustering algorithms rely on a number of arbitrary parameters. As mentioned above, there are also a number of other approaches, but these also suffer from the same problem. This creates problems when one is attempting to draw inferences from the output of a cluster analysis. To cope with a multiplicity of possible approaches and associated parameters, often in a situation without theoretical guidance, one is compelled to follow the suggestions of Bailey and Gatrell (1995, p. 294):

> Faced with the range of different clustering methods that are available, it is usually a good idea to use several methods in any particular application. If these different methods tend to agree broadly on group membership for a particular choice of the number of groups, then this is probably an indication that this grouping is reasonably

'robust' and not just an artifact of the particular method being used. Clustering methods will always produce groups; it is the responsibility of the analyst to determine whether these have any substantive meaning.

In fact this valuable advice could be extrapolated to data-mining methods in general. Data mining is essentially a search for patterns in data, but it can be very difficult to tell whether any output relates to genuine patterns in the data, or to some artefact of the data-mining technique being used. Unfortunately, the 'black box' nature of some approaches, where the internal workings of the method are intentionally hidden from the analyst, is hardly helpful in this situation. In our opinion, black box techniques are generally best avoided. If it is decided that data mining is appropriate in some situation, then some awareness of the internal workings of the methods used is essential when inferring whether the patterns detected are genuine. The remaining data-mining techniques discussed should be considered (and applied) in this context.

8.2.2.2 Neural networks A more recent advance in data mining is the use of neural networks (McCulloch and Pitts, 1943). The mechanisms behind neural networks are discussed in detail in many texts, for example Hertz et al. (1991) and Gurney (1995). Essentially, these are based on a 'machine learning' paradigm, and are loosely modelled on the 'real' neurons found in the brain. Briefly an (artificial) neuron can be described as a communications device whose input signals are combined to produce an output signal. In a basic network, when the weighted sum of the inputs exceeds some threshold level the output jumps from zero to one. This crude model can be extrapolated in a number of ways. For example, instead of a binary step function suggested above, a continuous logistic function could be substituted. In a neural network, a number of these neurons are connected together, typically in a number of layers as in Figure 8.2.

Presenting a network with a number of inputs will cause the first layer to produce a set of outputs (in this case there are five outputs from the first layer). These outputs will depend on the weights used in combining the inputs. These outputs are then fed as inputs to the next layer, which produces a new set of outputs which feed into the next layer, and so on until a final set of outputs (in this case two) are produced. If the weights are chosen well, each neuron will act as a 'feature detector', providing a high output when the inputs follow some particular pattern.[1] Thus, a five-neuron layer could respond to five distinct features. Having multiple layers allows one to detect 'features of features': for example, a second-layer neuron could detect when a particular pair of first-layer features are both present – a logical AND operation. If there are *m* inputs to the network, each input can be set to one variable in a data set, and the network can provide a 'response' to individual observations by reading values from the final output layer.

In typical neural computing applications, there are a lot of neurons, a lot of data and a lot of features to detect. In this sense, the approach can be thought of as a data-mining exercise. One wishes to use the layers in the network to detect

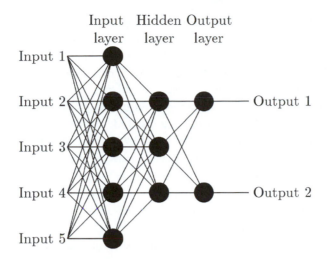

Figure 8.2 **A layered neural network. Neurons are solid circles; lines show connections linking outputs to inputs; outputs always flow to the right. The above network has five inputs and two outputs**

'features' (or patterns) in the data. The difficulty here is in choosing the weighting scheme and the threshold for each of the neurons. Each one of these can be thought of as a parameter in the network, and for a given data set one has to find suitable values to provide 'good' responses. One common approach, described as *supervised learning*, is to provide a training data set with a set of desired responses for each observation. Finding suitable weights and thresholds is then an optimization problem; weights must be found to make the network provide outputs which match the desired outputs as closely as possible. This optimization task is often referred to as *network training*. This is qualitatively similar to a statistical regression or discriminant analysis problem.

Unfortunately, the optimization task is often very difficult. For example, assuming each neuron has one weight per input, and one threshold, the network in Figure 8.2 has 56 parameters, and this network is much smaller than those generally used in practice. There are several possible approaches to the optimization task, but they are all computationally expensive. Several variations on this approach also exist – for example, some algorithms alter the configuration of the network, dynamically adding or removing neurons during the training process (Reed, 1993). Others replace the logistic or threshold response neurons with other having a different functionality – for example, the use of *radial basis functions* (Broomhead and Lowe, 1988).

An alternative approach to choosing weights is the notion of *unsupervised learning*. If supervised learning can be likened to discriminant analysis or regression, unsupervised learning parallels cluster analysis. When training an unsupervised network no set of desired responses is required. The network is left to find features in the data without the optimization-based guidance used in supervised networks. One approach is through competitive learning methods in which neurons providing the 'strongest' response (in terms of the highest output) to a particular input are encouraged to respond even more positively to this and similar inputs by adjusting weights appropriately. An example of this is found in *adaptive resonance theory* or ART (Carpenter and Grossberg, 1988). This, when stripped of its neural computing terminology, bears remarkable resemblance to *k*-means clustering as described in the previous section. Note that in unsupervised learning, the configuration of neurons often deviates from the layered formation usually found in supervised learning. For example, Kohonen (1989) arranges neurons in a lattice, and applies ART-like competitive training not only to the strongest responding neuron, but also to its topological neighbours within the lattice. As with supervised networks, typically networks are very large, and training them is a lengthy task. Feature identification is a key concept here, again justifying the concept of neural networks as a data-mining tool.

Inferentially, neural networks provide an interesting set of problems. First, supervised and unsupervised networks do slightly different jobs, and so inferences have to be drawn in different ways for the two methods. Supervised learning is essentially a predictive tool. Having trained a network to reproduce a desired set of outputs in training data, one then feeds in new data in the hope that the outputs produced will predict some (unobserved) real-life response to the data. Again, this is one way in which regression or discriminant analysis might be used. To make predictive inferences, however, one must consider the plausibility of the prediction mechanism as well as the prediction that the mechanism produces. This can cause problems. Some features detected in data might be genuine, and likely to provide useful information for prediction, but others might be spurious, coincidental phenomena not occurring in future data sets. These spurious features might be misleading, and could lead to erroneous predictions. In a neural network one needs to add enough neurons to detect 'genuine' features, but not so many as to detect the spurious features. Also one cannot be sure that the number of neurons does equal the number of genuine features and that each neuron will succesfully detect a genuine feature. This problem has parallels with the problem of model specification in conventional statistics. A network with too many neurons is similar to a regression model with too many variables. In the latter case, the inclusion of spurious variables tends to have an adverse effect on the prediction of response variables. Unfortunately, there is considerably less theoretical guidance for this kind of problem in the neural case than the statistical case.

Despite these theoretical difficulties, one claim made for neural networks is that their predictive performance exceeds that of 'statistical models'. This could perhaps be interpreted as providing an extra degree of confidence in the context of informal predictive inferences. However, this claim is not unchallenged:

> There is a paper by White at UCLA (*sic*) who tried using neural networks to forecast the closing price of IBM's stock. His paper, and a few others I have, all tend to conclude that using neural networks for such activities is no more accurate than using *well* traditional statistical techniques.

(G. Aharonian, taken from the newsgroup *comp.ai.neural-nets*, first reproduced in Barndorff-Nielsen et al. (1997). Here it is suggested that the 'White' referred to is probably Halbert White at UCSD – see White, 1988.

It is also suggested that many of the extravagant claims made on the performance of supervised neural networks are not usually made by their inventors (Watt, 1991).

Inferentially, the outlook for unsupervised neural networks is perhaps more positive. These are not used in a predictive context. As stated earlier, they are used in the manner of cluster analysis techniques. Indeed, one could consider using such techniques alongside conventional cluster analysis methods when considering a particular data set, and any statements about informal inference from cluster analysis could equally apply here. One added advantage that the Kohonen approach offers is the ability to visualize the organization of the neural network in a spatial way. To see this in detail, see either Kohonen (1989) or Gurney (1995).

8.3 Formal inference

Here probability must be mentioned for the first time. The inferential problem could be regarded as attempting to make a statement about unobservable information on the basis of linked observable information. For instance, one may wish to know whether an observed set of points are generated from a uniform (unclustered) distribution. The unobservable information here is whether the generating distribution is uniform or not, and the observable information is the list of point coordinates. In the last section, inferences relating to the unobservable information were made on an informal basis. However, in many instances it is helpful to provide some quantitative measure of the reliability of any statements relating to the unobservable information. This is typically dealt with using probabilities. Two common ways of providing probabilistic statements about unobservable information discussed here are *Bayesian* inference and *classical* inference. Both will be discussed here.

8.3.1 Bayesian inference

Bayesian inference is based on *Bayes' theorem* (Bayes, 1763), which relates conditional probabilities. Suppose $p(X)$ denotes the probability that statement X is true, and that $p(A|B)$ denotes the probability that statement A is true *given* that statement B is true. Then Bayes' theorem in its simplest form may be stated as

$$p(B|A) = \frac{p(A|B)p(B)}{p(A)} \qquad (8.1)$$

This can be quite simply derived; see Gelman et al. (1995), for example. In simple terms, Bayes' theorem relates the probability of A given B to the probability of B given A. This is extremely useful from an inferential viewpoint. Suppose A is the observable phenomenon and B the unobservable phenomenon. If one is able to say how likely an observed A is given B, one can use Bayes' theorem to make probabilistic statements about the unobservable B. To do this, the other quantities $p(A)$ and $p(B)$ should be considered. These are the *marginal probabilities* of A and B. $p(A)$ is the probability that A is true regardless of whether B is true or not; obviously $p(B)$ can be regarded in the same way with respect to A. $p(A)$ can be viewed in terms of the probability that A occurs if B is true and the probability that A occurs if B is not true:

$$p(A) = p(A|B)p(B) + p(A|\tilde{B})[1 - p(B)] \tag{8.2}$$

giving an alternative form of Bayes' theorem:

$$p(B|A) = \frac{p(A|B)p(B)}{p(A|B)p(B) + p(A|\tilde{B})[1 - p(B)]} \tag{8.3}$$

which is sometimes helpful. In the notation used here $\tilde{X}$ denotes the logical inversion of statement X – that is, 'not X'. The information one now requires to evaluate $p(B|A)$ is the probability of the observed data if B is true, the probability of the observed data if B is untrue, and $p(B)$. This last quantity is the marginal probability that the unobservable B is true regardless of whether A is true. In the Bayesian paradigm, this is regarded as the probability that B is true *prior* to observing A. This is perhaps best illustrated by a worked example.

Suppose you work in a store, and that all of the customers from this store come from either village X or village Y. In village X there are 100 residents, 30 of which own red cars, and 70 of which own non-red cars. In village Y there are 80 residents, 20 of which own red cars, and 60 own non-red cars. A red car draws up outside your store, but you cannot see the driver. What is the probability that the driver comes from village X, given the car colour?

Here, the unobservable information is whether the customer comes from village X or village Y, and the observed, linked information is that the customer has a red car. Thus, define statements A and B as follows:

- A: The customer has a red car.
- B: The customer lives in village X.

One can then determine the probabilities required for Bayes' theorem. $p(A|B)$ is the probability that the customer has a red car given he or she comes from village X. This is 30/100, or $\frac{3}{10}$. Similarly, $p(A|\tilde{B})$ is the probability that the customer has a red car given he or she does not come from village X. Since all customers come from either village X or village Y, this is the same as the probability of a customer owning a red car given he or she comes from village Y. This is $20/80$, or $\frac{1}{4}$. Finally, $p(B)$ is the probability that the customer comes from village X, regardless of the colour of the car. This is $100/(80+100)$ or $\frac{5}{9}$. One is then ready to compute $p(B|A)$, which was asked for in the initial question:

$$p(B|A) = \frac{p(A|B)p(B)}{p(A|B)p(B) + p(A|\tilde{B})[1 - p(B)]} = \frac{\frac{3}{10} \times \frac{5}{9}}{\frac{3}{10} \times \frac{5}{9} + \frac{1}{4} \times \frac{4}{9}} = \frac{6}{10}$$

Thus, after observing the colour of the customer's car, the probability that he or she comes from village X is $\frac{6}{10}$, or 0.6. Note that this computation could have been approached in a slightly different way, by evaluating $p(A)$ directly. $p(A)$ is just the probability that the customer has a red car regardless of the home village. Since there are 180 villagers in all, and the total number having red cars is $20 + 30 = 50$, then this probability is $\frac{50}{180}$ or $\frac{5}{18}$. Then, from the first form of Bayes' theorem,

$$p(B|A) = \frac{p(A|B)p(B)}{p(A)} = \frac{\frac{3}{10} \times \frac{5}{9}}{\frac{5}{18}} = \frac{6}{10}$$

Obviously, the result is the same as before.

This is a somewhat contrived example, but it contains all the ingredients of a Bayesian analysis. The probability that the customer comes from village X, $p(B)$, can be regarded as a prior probability for statement B. Before a car has arrived, this is the probability that the driver will come from village X. The probability that the driver comes from village X *after having seen the red car* is modified to $p(B|A)$. This conditional probability is usually referred to as the *posterior* probability of statement B.

In a more realistic Bayesian analysis, A is not a single observation but a collection of observations – in other words, an entire data set. Also, the unobservable information, B, is not usually a simple binary hypothesis. For example, instead of just two villages in the last example, there could have been m villages. If one calls these villages $1, 2, \ldots, m$, then let the events that the customer comes from each village be $B_1, B_2, \ldots, B_m$. In this case, Bayes' theorem takes the form

$$p(B_k|A) = \frac{p(A|B_k)p(B_k)}{p(A|B_1)p(B_1) + p(A|B_2)p(B_2) + \ldots + p(A|B_m)p(B_m)} \tag{8.4}$$

where k could be any village. This could be regarded as an attempt to make inferences about an unobservable *parameter* k on the basis of observed car colour. Since Bayesianism provides a formal approach, such inferences take the form of a probability distribution.

A further extension of the method allows inferences to be made regarding *continuous* parameters. In this case, alter the notation by replacing B with an unobserved vector of parameters, θ, and replacing A with an observed vector of observations $\mathbf{x}$. This relates to more usual statistical problems. For example, in a regression model $\mathbf{x}$ would represent observed data, and θ would represent unobserved regression coefficients. In this situation, the summation in the discrete equation (8.4) is replaced by an integral:

$$p(\theta|\mathbf{x}) = \frac{p(\theta)p(\mathbf{x}|\theta)}{\int p(\theta)p(\mathbf{x}|\theta)\,d\theta} \tag{8.5}$$

The above is perhaps the most common form of Bayesian inference. For example, θ might be the mean and standard deviation of a normal distribution, say (μ, σ), and $\mathbf{x}$ a set of observations. Then $p(\mathbf{x}|\theta)$ would just be the multivariate normal probability density function, with each element of $\mathbf{x}$ having mean μ, and the variance–covariance matrix being $\sigma^2\mathbf{I}$. Thus

$$p(\mu, \sigma|\mathbf{x}) = \frac{1}{(2\pi\sigma^2)^{n/2}} \prod_{i=1}^{i=n} \exp[-(x_i - \mu)^2/2\sigma^2] \tag{8.6}$$

Here, $p(\mu, \sigma)$ is harder to consider. In the simple example, the prior probabilities could be calculated from empirical knowledge. However, there is no clear guidance here as to the prior probability function. This is addressed in a Bayesian framework by asserting that $p(\theta)$ represents the analyst's prior *beliefs* about θ. If the analyst has no prior beliefs about θ, one may represent this by setting $p(\theta)$ to be a uniform distribution. When this is done, $p(\theta)$ is referred to as a *non-informative prior*.

Unfortunately, this will run into problems when θ can take on an infinite range. In the normal distribution example, μ could range from $-\infty$ to ∞. A non-informative prior in this case is not a proper probability distribution, since a constant value does not have a well-defined integral between $-\infty$ and ∞. However, this need not be a problem, provided the posterior distribution is still well defined. Again, in the above example, if one does adopt a non-informative prior for μ, say $p(\mu) = 1$, and assume σ is known, then the posterior distribution is just

$$p(\mu|\mathbf{x}) = \frac{(2\pi\sigma^2)^{-n/2} \prod_{i=1}^{i=n} \exp[-(x_i - \mu)^2/2\sigma^2]}{(2\pi\sigma^2)^{-n/2} \int_{-\infty}^{\infty} \prod_{i=1}^{i=n} \exp[-(x_i - \mu)^2/2\sigma^2]\,d\mu} \tag{8.7}$$

It can be verified that $p(\mu|\mathbf{x})$ is itself a normal distribution with mean $\bar{x}$, the sample mean of $\mathbf{x}$, and variance σ^2/n. Typically, σ would not be known, and one would need to consider the joint posterior distribution of μ *and* σ (Gelman et al., 1995).

Bayesian methods may also be applied to more complex spatial models. For example, Diggle et al. (1998) consider a Bayesian analysis of kriging (Isaaks and Srivastava, 1988), a spatial regression technique considered elsewhere in this book. In this case, the observed data are a set of (x, y, z) points, and the unknown parameters are the kriging coefficients together with the regression coefficients. This is a good example of 'real-world' Bayesian analysis, and highlights several practical difficulties. One of the most notable problems is the fact that the integral in the denominator in Bayes' theorem is often analytically intractable. In more realistic terms, Bayes' theorem has to be stated as

$$p(\theta|\mathbf{x}) \propto p(\theta)p(\mathbf{x}|\theta) \tag{8.8}$$

where the constant of proportionality is generally unknown. All that is known is that it causes the right hand side of (8.8) to integrate to unity. How can one understand the properties of a posterior distribution if one cannot derive the distribution analytically? One approach to this is through simulation. If it is possible to simulate θ values drawn from $p(\theta|\mathbf{x})$ then one can produce large samples of such values, and investigate the posterior distributional properties of θ through these samples. Furthermore, using the Metropolis algorithm (Metropolis and Ulam, 1949; Metropolis et al., 1949) it is possible to simulate $p(\theta|\mathbf{x})$ without knowing the proportionality constant. These two facts, together with an abundance of personal computers that are powerful enough to generate such simulations reasonably quickly, have led to a widespread adoption of Bayesian methods in recent years.

Thus, Bayesian inference allows us to make statements about unobserved parameters in terms of probability distributions. It is also possible, in some cases, to compare models. One way of doing this is to consider *Bayes factors*. Suppose H_1 and H_2 are two competing models. For example, H_1 might be an ordinary regression model, and H_2 might be the same regression problem with spatially correlated errors. The ratio of the posterior likelihoods can be expressed as

$$\frac{p(H_1|\mathbf{x})}{p(H_2|\mathbf{x})} = \frac{p(H_1)}{p(H_2)} \times \text{Bayes factor } (H_2, H_1) \tag{8.9}$$

Thus, starting off with prior odds of H_1 to H_2 and then multiplying by the Bayes factor gives the posterior odds of H_1 to H_2. Thus, the Bayes factor is an *odds ratio* showing how the odds of one model against another are altered in the light of the data $\mathbf{x}$. Thus, after a typical analysis one might draw a conclusion such as: 'having collected data it seems that the odds *spatial regression model:ordinary regression model* are now four times greater'. Note that the equations above refer to models, not parameters. It is usual that the models contain many parameters in common, but not essential. For example, the same approach could be used to compare a spatial regression model with a Bayesian GWR model.

However, this methodology is not without controversy. Bayesian inference requires a set of observations, a probabilistic model linking the unobserved parameters to the observations, and a prior distribution for the unobserved parameters. The last item required is the source of some debate. As identified earlier, a state of no knowledge is represented by a non-informative prior. However, some Bayesians argue that in many cases, this is an unrealistic representation of prior knowledge. For example, there may have been previous studies attempting to estimate θ. In this case, if a previous study was subjected to Bayesian analysis, its posterior distribution could be used as a prior for the new study. Others take a stronger viewpoint: one can have subjective prior opinions about the value of θ, possibly due to qualitative knowledge.

For example, in a study to estimate θ, the average distance that people travel to work in the UK, it could be argued that a non-informative improper prior for θ in the range 0 to ∞ should be used. However, this is stating that θ is equally likely to be 10 miles or 10 million miles. Subjectively, this seems nonsensical. A more reasonable approach might be to use a proper uniform prior in the range 0 to 200 miles, or perhaps a non-uniform prior distribution where the higher values of θ are less likely, but not impossible. Here an exponential distribution might be appropriate.

There are others who find arguments of the kind given above to be unpalatable. Essentially, this approach requires probabilities to be used in a subjective sense, which some argue to be meaningless. It could be argued that informative priors, if used without constraint, could lead analysts to draw any conclusion they wish from experimental data. It is difficult to answer the above objection, but it can be argued (see next section) that classical approaches to statistical inference suffer from similar problems, although they are more subtly hidden. On a more optimistic note, it can be demonstrated that, provided the prior distribution follows some mild regularity conditions, the posterior distribution becomes more and more concentrated around the true value of θ as the sample size for $\mathbf{x}$ (i.e. n) tends to ∞ – see for example Gelman et al. (1995). That is, regardless of our choice of prior,[2] as more and more data arrive, the posterior distribution will get closer and closer to a single mass point at the true value of θ – with enough data even very poor priors do not affect the outcome too much.

8.3.2   Classical inference

Classical inference is probably the most familiar kind of statistical reasoning for geographers. This is the inferential system which encompasses significance tests and confidence intervals – much of spatial statistical analysis makes use of these concepts. Despite this, it is often argued that classical inference is misunderstood by many students (Berry, 1997). In this section, it is intended to outline the basic principles and compare them with the Bayesian concepts introduced in Section 8.3.1.

Perhaps surprisingly, Berry (1997) states that many students who are taught only classical inference misunderstand it in a way which coincides with the definition of Bayesian inference! However, in the context of hypothesis testing, the difference between the Bayesian and the classical approach is very clear. Regardless of whether probabilities are subjective or objective, a Bayesian draws conclusions of the form 'the probability that hypothesis X is true is y'. In classical inference, hypothesis X (the *null hypothesis*) is not regarded in probabilistic terms. It is assumed to be a state of nature which is either true or false. Here, only the data are regarded as probabilistic. Classical inference then goes on to consider questions of the form 'how likely is it to obtain this particular data set *given* a particular hypothesis'. To determine the truth or falsehood of the null hypothesis, a numerical function (often called the *test statistic*) is applied to the data. This is itself a random function, and the classical test then reduces to the question 'how likely is it to obtain a test statistic at least as extreme as that observed given a particular hypothesis?'. Testing whether this statistic exceeds a particular value provides an answer of true or false to this question – but of course it could be wrong. One can then go on to consider the probability that the answer is wrong. There are essentially four possible outcomes as shown in Table 8.2.

The two probabilities of interest are then α, the probability that the test is wrong when the null hypothesis is true, and $1 - \beta$, the probability that the test is wrong when the null hypothesis is false. α is referred to as the *significance level* of the test, β as the *power* of the test. Thus, the probabilities in classical inference do not refer to the truth of the null hypothesis, but to the success of the testing process. They are *operating characteristics* of the test. The significance level of the test is fairly easy to specify, provided one can specify the probability of the test outcome in terms of the null hypothesis. There are a vast number of examples of tests of this kind in many standard statistical texts – typical are tests of whether a particular model parameter is equal to zero.

The power is generally harder to specify. Typically, the null hypothesis is a simple mathematical statement – such as a particular regression coefficient being zero. However, there are an infinite number of alternatives to this: that is, any value of the previous regression coefficient *except* zero. Clearly, the power of the test will

Table 8.2 **Probabilities used in classical inference: α is the probability that the result of the hypothesis test is 'false' when the null hypothesis is true. β is the probability that the result of the hypothesis test is 'false' given the null hypothesis is false**

		Test result	
		True	False
Null hypothesis	True	$1 - \alpha$	α
	False	$1 - \beta$	β

depend on how different from zero the regression coefficient is. Thus, investigations into the power of a test often involve plotting a function of the true value of the coefficient against the power. However, this could be argued to be too narrow a viewpoint. A hypothesis could be wrong owing to a completely misspecified model, not simply a model of almost identical form in which one coefficient deviates from zero.

This classical approach can also be applied to parameter estimation. In this case, a numerical hypothesis test is replaced by two functions applied to the observed data **x**, giving an upper and lower bound for some parameter θ, called a *confidence interval*. In this case, one is concerned with α, the probability that θ will be contained within the confidence interval. Again, α is an operating characteristic of the estimation process, *not* the probability that θ lies within the confidence interval. θ is assumed to have a fixed value, and inference is made in terms of the chance of the confidence interval containing θ. A typical example of this is found in the $p\%$ confidence interval for the mean of a normal distribution, given a sample vector **x**:

$$\bar{x} \pm t_{n-2,\,p/200} \frac{s}{\sqrt{n}} \tag{8.10}$$

where n is the sample size, $\bar{x}$ is the sample mean, s is the sample standard deviation, and $t_{a,b/100}$ is the bth percentile of the t-distribution with a degrees of freedom.

Maximum likelihood estimators also play an important role in classical inference. The *likelihood* of **x** given θ is $p(x|\theta)$. Then, considering **x** as fixed, the likelihood is a function of θ, say $l(\theta)$. Then, the maximum likelihood estimate of θ, usually written $\hat{\theta}$, is the value of θ maximizing $l(\theta)$. Furthermore, it can be shown that if $L(\theta) = \log{(l(\theta))}$, then as the sample size tends to infinity, the sampling distribution of $\hat{\theta}$ tends to a normal distribution with mean θ and variance $-1/L''(\theta)$ (DeGroot, 1988). As with the Bayesian proof of asymptotic consistency, mild regularity conditions are required for $l(\theta)$. Again, if n is reasonably large then $\hat{\theta}$ is a good approximation for θ, and an approximate confidence interval for θ is

$$\hat{\theta} \mp \frac{N_{p/200}}{L''(\hat{\theta})} \tag{8.11}$$

where $N_{a/100}$ denotes the ath percentile of the normal distribution. This can be extended to multidimensional θ values.

Thus, the classical approach provides a formal inferential framework for both hypothesis testing and parameter estimation. However, there are some objections to this methodology. Most of these target hypothesis testing. The main problem here is that most null hypotheses can be expressed in the form $\theta = 0$ for some parameter θ in a model. Suppose, however, that in reality θ was not actually zero, but some small number very close to zero, say $0.000\,01$. If n were sufficiently large, a classical hypothesis test would reject the null hypothesis of $\theta = 0$. However, although this would be a significant result in classical terms, would it be of any real interest? The problem arises because the null hypothesis states that θ is a point of zero measure on the real line. Any infinitesimal deviation from this contradicts the hypothesis, and given enough data, the null hypothesis is likely to be rejected. As early as 1957, this problem was identified:

> Null hypotheses of no difference[3] are usually known to be false before the data are collected ... when [the data are] collected their rejection or acceptance simply reflects the size of the sample and the power of the test, and is not a contribution to science.
>
> (Savage, 1957)

Nearly four decades later, strong objections are still being raised:

> Continued association with hypothesis testing is not in our own best interest. I believe that statisticians would be unwise to seek the limelight in any forthcoming 75th anniversary, centennial or tricentennial celebrations of hypothesis testing.
>
> (Nester, 1996)

The problem seems to be that it is not sensible to ask 'is θ exactly zero?'. A more sensible question might be 'is θ sufficiently different from zero to matter?'. The last question has to be answered in context. For example, in a regression model, say $E(y) = a + bx$, b might not be exactly zero, but one could consider whether it is large enough to cause substantial variation in $E(y)$ over a range along which x might be reasonably expected to vary. Nester (1996) argues that the problem can be avoided by switching the emphasis of classical inference away from hypothesis tests, and concentrating solely on parameter estimation via confidence intervals. There appears to be some justification in this. Instead of testing whether a parameter has a specific value, one attempts to obtain an interval estimate of the parameter. The size of the interval gives some idea of the certainty of the statement about the parameter value, and looking at the magnitudes of the upper and lower interval bounds, one can gain some impression of whether the parameter is sufficiently large to be of consequence.

8.3.3 Experimental and computational inference

8.3.3.1 Classical statistical inference and spatial autocorrelation Classical statistical inference operates as follows. A null hypothesis is stated, such as 'the population from which this sample was drawn has a parameter value of zero', and a statistic, such as a t-statistic or a z-score, is calculated from the sampled data set and then compared with a theoretical distribution with known probability properties (e.g. a normal distribution). On the basis of this comparison we can reject or accept the null hypothesis according to some a priori and arbitrary cut-off point and we can state the probability that if we accept the null hypothesis that our conclusion is

incorrect. Alternatively, and often preferably, we can derive a confidence interval
for the parameter which consists of a range of values within which the unknown
population value of the parameter lies with a stated degree of confidence. Whichever approach we take to making inferences using this classical approach, it is
necessary to be able to assume some form of theoretical distribution for the test
statistic. For some statistics, such as the sample mean and OLS parameter
estimates, the theoretical distributions are well-known and can, in most circumstances, be used with confidence that the assumptions concerning the distributions
are met. However, for some statistics, either there is no known theoretical distribution against which to compare the observed value, or, where the distribution is
known, the assumptions underlying the use of that particular distribution are
unlikely to be met. Both of these circumstances are common in the analysis of
spatial data and here the construction of experimental distributions becomes
especially useful (Diaconis and Efron, 1983; Mooney and Duvall, 1993; Efron and
Gong, 1983; Efron and Tibshirani, 1986).

The central idea in the use of experimental distributions for statistical inference
is that the sampled data can yield a better estimate of the underlying distribution of
the calculated statistic than making perhaps unrealistic assumptions about the
population. The sampled data are resampled in some way to create a set of samples,
each of which yields an estimate of a particular statistic. If this is done many times,
the frequency distribution of the statistic forms the experimental distribution
against which the value from the original sample can be compared. Consequently,
experimental distributions can be constructed for any statistic, even if the theoretical distribution is unknown – as is the case, for example, with statistics such as
the mean nearest-neighbour distance for an irregularly shaped study area.

As an example of the construction of experimental distributions for spatial data,
consider the calculation of the spatial autocorrelation coefficient, Moran's I,
described in Chapter 1, the formula for which is

$$I = \left(\frac{n}{\sum_i \sum_j w_{ij}}\right) \left(\frac{\sum_i \sum_j w_{ij}(x_i - \bar{x})(x_j - \bar{x})}{\sum_i (x_i - \bar{x})^2}\right) \quad (8.12)$$

where i and j index the spatial units of which there are n, $\bar{x}$ is the mean of x and w_{ij}
is the weight (degree of connection) between zones i and j. The weights are usually
determined either by continuous inverse distance measurements or by binary
definitions of whether or not the two zones are contiguous. In the discussion and
examples which follow, we will assume that weights are defined by the more
realistic inverse distance measure. The expected value of Moran's I (i.e. the value
that would be obtained if there were no spatial pattern to the data) is

$$E(I) = -\frac{1}{n-1} \quad (8.13)$$

with values of I larger than this indicating positive spatial autocorrelation (similar values cluster together) and values below this indicating negative spatial autocorrelation (similar values are dispersed).

As mentioned in Chapter 1, there are two theoretical formulae which can be used to calculate the variance of I assuming that the weights are defined as continuous variables. One of the formulae assumes that each observed value of the attribute x is drawn independently from a normal distribution, in which case the variance of I is given by

$$\text{var}(I) = \frac{n^2 S_1 + n S_2 + 3(\sum_i \sum_j w_{ij})^2}{(\sum_i \sum_j w_{ij})^2(n^2 - 1)} \tag{8.14}$$

where

$$S_1 = \frac{\sum_i \sum_j (w_{ij} + w_{ji})^2}{2} \tag{8.15}$$

and

$$S_2 = \sum_i \left(\sum_j w_{ij} + \sum_j w_{ji} \right)^2 \tag{8.16}$$

The other results from the assumption that the process producing the observed data pattern is random and the observed pattern is just one out of the many possible permutations of n data values distributed in n spatial units. Under this assumption, the variance of I is given by

$$\text{var}(I) = \frac{n S_4 - S_3 S_5}{(n-1)(n-2)(n-3)(\sum_i \sum_j w_{ij})^2} \tag{8.17}$$

where

$$S_3 = \frac{n^{-1} \sum_i (x_i - \bar{x})^4}{(n^{-1} \sum_i (x_i - \bar{x})^2)^2} \tag{8.18}$$

and

$$S_4 = (n^2 - 3n + 3)S_1 - nS_2 + 3\left(\sum_i \sum_j w_{ij}\right)^2 \qquad (8.19)$$

and finally

$$S_5 = S_1 - 2nS_1 + 6\left(\sum_i \sum_j w_{ij}\right)^2 \qquad (8.20)$$

The distribution of I is asymptotically normal under either assumption (normality or randomization); in other words, as long as n is 'large', the following standardized statistic can be calculated,

$$Z = \frac{I - E(I)}{\text{var(I)}} \qquad (8.21)$$

and reference made to normal probability tables. The questions raised by this procedure are 'how large does n have to be?' and 'how well does the assumption of asymptotic normality hold even if n is large?'. Here we compare the conclusions reached from this classical use of statistical inference with those obtained from using an experimental distribution for I.

8.3.3.2 Experimental distributions and spatial autocorrelation Suppose we have n values of an attribute distributed across n spatial units and that we have calculated Moran's I for this distribution with a given definition of spatial weights. In terms of statistical inference, the basic question we need to answer is: 'What is the probability that a spatial pattern as extreme as the one observed could have arisen by chance?'. This can be answered experimentally by permuting the n attribute values across the n spatial units many times and each time calculating the value of Moran's I until an experimental distribution for I is constructed. The specific steps are as follows:

1   Calculate I for the observed distribution of the attribute x and call this I^*.
2   Randomly reassign the n data values across the n spatial units.
3   Calculate I for the new spatial distribution of the attribute x and store.
4   Repeat steps 2 and 3 many times (at least 99 times and preferably 999 times).

This will produce an experimental distribution for I against which the value of I^* can be assessed. The proportion of values in the experimental distribution which equal or exceed I^* yields an estimate of the probability that a value of Moran's I as high as I^* could have arisen by chance.

The experimental distribution can also be used to create a confidence interval for I around I^*. In the usual manner, an α-level confidence interval for I is defined as the range of values that will include the true value of I with probability $(1 - \alpha)$. Usually this is centred around the estimated value of I for the observed data, I^*. To obtain this interval it is clearly necessary to assume something about the variability of the distribution being examined. For instance, in classical statistical inference, it is common to assume that the distribution is normal and then the $\alpha\%$ level confidence interval for I would be

$$I^* - z_{\alpha/2}\sigma_I < I < I^* + z_{\alpha/2}\sigma_I \qquad (8.22)$$

where $z_{\alpha/2}$ represents the value of z in a normal probability table corresponding to a probability of making a Type I error of α. The value of σ_I is estimated from either Equation (8.14) or (8.17). The experimental distribution can also, of course, yield an estimate of σ_I although there are several ways in which this can be achieved (Mooney and Duvall, 1993). Here we discuss two of them.

One is very similar to the above with the estimate of σ_I derived from the experimental distribution by

$$\hat{\sigma}_I = \left(\frac{\sum_{k=1,m}(I_k - \bar{I})^2}{m - 1} \right)^{1/2} \qquad (8.23)$$

where k indexes the calculated value of I for a particular random assignment of the data to the n spatial units, $\bar{I}$ represents the mean of these values and m indicates the number of times I_k is calculated. This estimate of σ_I is inserted into (8.22) along with the points of the z-distribution associated with probabilities $\alpha/2$ and $1 - \alpha/2$.

The confidence interval specified in Equation (8.22) assumes the distribution of I is normal and centred on I^* whereas in fact it will be centred around the expected value of I, as given in Equation (8.13). An alternative therefore is that provided by Bickel and Freedman (1981) and Efron (1981) and referred to as the percentile-t method. In this, each value in the experimental distribution, I_k, is transformed into a standardized variable, t_k:

$$t_k = \frac{I_k - \bar{I}}{\sigma_I} \qquad (8.24)$$

where σ_I can be estimated either analytically using (8.14) or (8.17) or from the experimental distribution using (8.23). The standard error is a constant for all calculations of I since the value depends solely on the spatial arrangements of the zones for which the data are reported (see Equations (8.14) and (8.17)). This conveniently removes the need for a further set of empirical distributions to be constructed to estimate individual standard errors as suggested on page 41 of Mooney and Duvall (1993).

Application of Equation (8.24) to each experimental value of I results in an experimental distribution for t_k from which the $\alpha/2$ and $1 - \alpha/2$ percentile values can be obtained to develop a confidence interval for I around $\bar{I}$ as

$$\text{prob}[\bar{I} - t_{\alpha/2}\sigma_I < I < \bar{I} + t_{1-\alpha/2}] = 1 - \alpha \tag{8.25}$$

There is evidence to suggest that the percentile-t method is the more accurate (Hall, 1988; Hinckley, 1988; Loh and Wu, 1987) but ultimately the choice of method is up to the analyst and might not be critical in many situations. Below, we compare both experimental methods of computing a confidence interval for I with the classical procedure using both Equations (8.14) and (8.17) to compute the standard error.

8.3.3.3 An empirical comparison of classical and experimental inference   To consider the role of both classical and experimental inference techniques in quantitative geography, an example based on Moran's I-statistic will now be given. For this example, a study region (corresponding to an inner city and some of its immediate suburbs) of 26 spatial units was chosen, and for each of these, the number of *economically active people* (i.e. those capable of working) and the number of people seeking work were obtained. By expressing the latter as a percentage of the former, a rate of unemployment was computed for each area. These are mapped in Figure 8.3.

Moran's I coefficient of spatial autocorrelation for these data gives an indication of the degree to which areas of high and low unemployment tend to cluster together. To compute this coefficient, we need to define a spatial association matrix, **W**. In this case, this is a 26 by 26 square matrix describing the proximity of each pair of the areas. Here, we use $w_{ij} = \exp(-d_{ij})$, where d_{ij} is the distance between the centroids of area i and area j, in kilometres. Using Equation (8.12), Moran's I coefficient can be computed for the unemployment rates defined above. Here, we obtain $I = 0.266$. This is positive, suggesting that areas of similar unemployment levels do tend to cluster together to some extent.[4]

However, two questions arise from this computation:

1   Is this value of I notably different from the $E(I)$ for a sample from a process with no spatial autocorrelation?
2   How reliable is this estimate of I?

Computing confidence intervals for I would help to answer both of these questions. First, one can check whether both the upper and lower α confidence bounds for I are greater than zero – this is equivalent to an α-level significance test of the hypothesis that $I = 0$. Secondly, inspecting the width of the α-level confidence

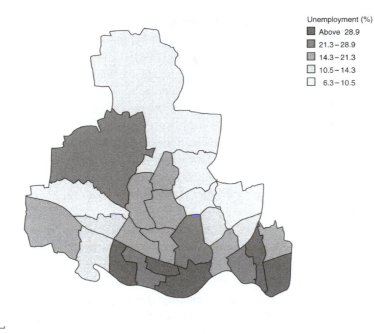

Unemployment (%)
■ Above 28.9
■ 21.3 – 28.9
■ 14.3 – 21.3
□ 10.5 – 14.3
□ 6.3 – 10.5

2 km

Figure 8.3   **Spatial distribution of unemployment rates**

interval, one gains a practical idea of the accuracy of the estimate for I. Given some of the objections to significance tests documented elsewhere in this chapter, the latter exercise would be regarded by some to be the more valuable.

Here, we will compute 5% confidence intervals for I using four alternative methods, two based on the classical approach and two making use of computer simulation. The classical approaches estimate the sample variance of I under the null hypotheses of an independently normally distributed unemployment rate (Equation (8.14)), and a randomization hypothesis (Equation (8.17)). The difficulty here is that in each case a set of assumptions is made about the distribution of the unemployment rate and about the asymptotic properties of the estimator for I. We cannot be certain that the distributional assumptions hold, and it is also doubtful whether asymptotic distributional properties can reasonably be invoked when analysing a data set with only 26 observations. For these reasons we will also analyse the data using experimental techniques, which substitute computer simulations of the exact distributions for the classical approximations.

The two experimental techniques both consider the randomization distribution of I, that is the distribution of I obtained when unemployment rates are randomly 'shuffled' between areas. The null hypothesis here is the assumption that owing to a lack of spatial pattern, any matching of the 26 unemployment rates to the 26 areas is equally probable. Here, an experimental sampling distribution of I is obtained by carrying out 9999 random 'shuffles' and computing I for each one. From this, two experimental confidence intervals can be computed, the first assuming that the sampling distribution of I is still normal[5] but computing the sample variance from the experimental distribution (Equation (8.22)). The second approach directly computes the interval from percentile points of the standardized experimental distribution – the so-called percentile-t-approach.

The results of the four methods of constructing confidence intervals are set out in Table 8.3. From these it can be seen that in general, intervals are wider for the classical methods than for experimental ones. The classical assumptions appear to have provided somewhat larger standard errors for I than those obtained from simulation. In this case, the experimental approach has provided us with a tighter set of 95% confidence intervals than classical techniques, although this does not always happen.

Using probability density estimation techniques such as those described in Chapter 4, it is possible to draw the sampling distribution probability density curve for I based on the experimental distribution. This is done in Figure 8.4. From this figure it is clear than the distribution is slightly skewed to the right, suggesting that the approximation of asymptotic normality is not appropriate here. This suggests that the most appropriate confidence interval here is the experimental percentile-t-based method. Note that this estimate has a notably higher value for the lower 95% confidence bound than the other three.

It is worth noting that although all four sets of confidence limits exclude zero, giving strong evidence that there is some autocorrelation in the spatial patterns of unemployment rates, the confidence intervals are really quite wide. This can be interpreted as noting that with just 26 observations, it is difficult to obtain a very reliable measure of Moran's I. In short, although we can be fairly certain that I exceeds zero, we do not have a clear picture of its exact value.

This exercise has shown how some extra insight into the sampling distribution of certain statistics can be gained by experimental approaches. It is also useful to consider how many simulations are necessary to obtain a reliable estimate of the sampling distribution for I. Here 9999 simulations were used to obtain the distribution estimate used in the computations above and seen in Figure 8.4. However, in Figure 8.5, distribution estimates for 19, 99, 499 and 999 experiments are also shown.

Table 8.3 Classical and experimental 5% confidence intervals for I. Note that $I_{0.025}$ and $I_{0.975}$ denote the 2.5th and 97.5th percentiles for I respectively

		I	$E(I)$	$SE(I)$	$I_{0.025}$	$I_{0.975}$
Theoretical	Normal	0.266	−0.038	0.118	0.034	0.497
	Randomized	0.266	−0.038	0.121	0.028	0.504
Experimental	Normal	0.266	−0.036	0.095	0.079	0.453
	Percentile-t	0.266	−0.036	0.095	0.107	0.483

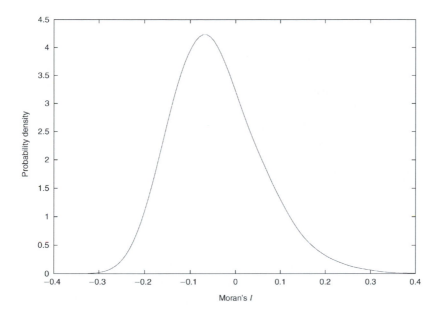

Figure 8.4 **Experimental sampling density for *I***

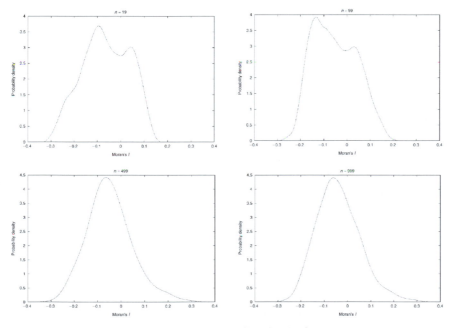

Figure 8.5 **Evolution of the experimental sampling density for *I***

From these it may be seen that for $n = 19$ and $n = 99$ simulations, the sampling distributions are quite erratic; by the time 499 simulations have been carried out, the result is very similar to that for 9999 simulations. This suggests that, in this case, 499 simulations would have been sufficient.

8.3.4 Model building and model testing

In both Bayesian and classical approaches to formal inference, an important goal is that of model specification. Essentially, one is using observed data to decide which theoretical model is most appropriate. The two formal approaches here offer quite different approaches to this. A very typical approach in classical inference is to consider a number of parameters and apply significance tests to each one in turn. This is particularly relevant in the case of, say, multiple regression, where a number of regression coefficients are tested to see whether they differ from zero. A number of problems have to be addressed here. Putting aside the objections to hypothesis testing raised in the last section, some new problems arise when more than one hypothesis is tested at a time. This is particularly true if a very large number of tests are taking place. Suppose 40 tests of the $\theta = 0$ type on regression coefficients are carried out with a significance level of 5%. On average, one would expect two of them to give a false rejection of the hypothesis even if the coefficients were all zero in actuality – assuming all of the tests are independent. Suppose, however, that tests said that four coefficients were non-zero. How easy would it be to decide which, if any, of the four tests were linked to genuinely non-zero coefficients?

The problem is compounded if one can no longer assume independence between the tests. Indeed, much statistical literature is concerned with the problem of *collinearity* in regression models; a phenomenon in which several of the independent variables are correlated. This results in non-independent coefficient tests. The problem here is that a spurious false rejection of the null hypothesis on one coefficient is likely to be accompanied by false rejections of null hypotheses associated with correlated coefficients. Another associated problem is that omitting a variable from a model (or adding one to it) can change the outcome of a test for another coefficient. All of this can make multiple significance tests in large models extremely difficult to interpret.

A number of ways of addressing this problem have been considered. A useful summary is given by O'Neill and Wetherill (1971), although others have argued that such techniques are the product of misplaced ideals:

> Multiple comparison methods have no place at all in the interpretation of data.
>
> (Nelder, 1971)

One of the most plausible approaches is to consider *all* of the parameters of interest simultaneously. Suppose there are m parameters. Then, instead of considering confidence *intervals* for each individual parameter one considers a confidence *region* in m-dimensional space. As before, the probability of this region containing the true value of the m-dimensional point is α. For example, in two dimensions one might have a confidence circle, or a confidence ellipse. This approach is helpful: by specifying multidimensional regions one can express the interdependence of the parameter estimates. A similar approach is possible in Bayesian inference, using multivariate posterior distributions. In this case, the probability that the estimates lie in a given region is computed. Interestingly, this method of resolving the problem moves away from hypothesis testing and towards parameter estimation, in accordance with some of the arguments of the previous section. Some of these ideas can also be applied to more general models, where it is not coefficients but arbitrary functions that are being modelled (Hastie and Tibshirani, 1990). This latter approach then becomes very similar in functionality to neural networks, but with a formal inferential framework.

It is worth noting that in both Bayesian and classical inference there are a number of other approaches to model checking and model specification, most of which are the subject of some debate. For some examples see Mallows (1973) for a classical perspective, or Gelman et al. (1995) for a Bayesian viewpoint.

8.4   Conclusions

The purpose of this chapter has been twofold. First, it was intended to showcase a number of different approaches to inference based on collected data. Secondly, it was intended to show that there is a great deal of current debate on the merits or otherwise of some of these techniques. At first glance this may seem to be a matter solely for statisticians, and of little concern to geographers. However, the debate relates to research methods often used by geographers. Perhaps a geographer using a significance test at some stage in his or her research would do well to consider that this is not a technique that is universally accepted as a justifiable approach – not just by the detractors of quantitative methods but also by some of their practitioners.

So what lessons should be learned from this? First, there is no such thing as 'the statistical approach'. For every data set and every research hypothesis there are always several possible quantitative approaches to choose from. 'Statistics' and 'statistical inference' are really umbrella terms for an ever increasing collection of distinct methodologies and the debates surrounding them. A practical question is then 'which techniques should one adopt?'. The advice here is not to be dogmatic. Each of the inferential systems described here is a model for reasoning in the presence of empirical data, and depending on context certain models may work better than others. For example, when one has a very clearly stated hypothesis arising from theory, then one of the more formal approaches is best, possibly using the experimental methodology if the underlying theory is mathematically complex. However, without such clear questions an exploratory approach may be more useful. A practical standpoint is to adopt the most suitable approach for a given problem, but at the same time not to be blind to any debate surrounding that approach.

Notes

1. If the neurons use weighted sums of inputs, it is only linear features (such as a two-dimensional input lying above a line in the plane) which may be detected.

2. Almost! Remember the 'mild regularity conditions'.
3. Essentially these are the same as null hypotheses of $\theta = 0$ if θ is a difference between parameters.
4. The largest possible value of Moran's I is 1.
5. Doubtful unless $n = 26$ is a sufficiently large sample for approximately asymptotic properties to hold.

9 Spatial Modelling and the Evolution of *Spatial* Theory

9.1 Introduction

A theme introduced in Chapter 1 is that quantitative geography has recently reached a stage of maturation whereby it is no longer a net importer of ideas and techniques from other disciplines but, rather, it is a net exporter, particularly in the areas of spatial statistics and geographical information science. The importation of ideas from other disciplines is not in itself to be discouraged and there is clearly a great deal to be gained by such cross-fertilization. However, since the beginnings of quantitative geography in the late 1950s and early 1960s and lasting until quite recently, there has been an unhealthy reliance on the use of statistical techniques and mathematical models developed in predominantly aspatial disciplines such as economics and physics. For instance, quantitative geographers have flirted with topics such as entropy (and appeals to the second law of thermodynamics!), portfolio analysis, Fourier analysis, neurocomputing, and new urban economics, all of which could be argued to have rather limited applicability to *spatial* processes concerning human decision making.

Concomitant with this overreliance on techniques developed in other disciplines, there has been a relative dearth in the development of statistical techniques and mathematical models aimed at understanding purely spatial processes. Consequently, a great deal of criticism, much of it reasonable, some of it not, has been levelled at modelling techniques which have been applied to human spatial behaviour but which make assumptions such as homogeneity of individuals, spatial non-stationarity, omniscience, rationality, equilibrium and optimal behaviour. Such criticism has at times been very powerful and influential (Sayer, 1976; 1992) with the result that there has been a decline of research into spatial modelling. Even the frameworks provided by classic spatial models such as central place theory (Christaller, 1933) and Weber's location theory (Weber, 1909) have been discarded by many geographers because of the unreasonableness of some of the assumptions embedded in them.

The criticism of certain areas of quantitative geography is not necessarily a bad thing. Some areas of research will undoubtedly, with hindsight, be seen as evolutionary dead-ends and the discarding of certain methodologies or topics is an inevitable part of the evolution and maturity of a discipline or subdiscipline. Who knows what parts of geographical enquiry, currently popular, will be seen as the academic equivalent of the dinosaurs in 20, or even 10, years' time? However, there are times when the criticism of spatial modelling has had an unwarranted negative effect. For instance, some frameworks, such as spatial interaction modelling, still get criticized for what they once were rather than for what they now are. Sometimes the initial criticisms, albeit reasonable, have led to calls for the complete termination of research into a particular modelling topic rather than for more research to correct what have turned out to be fairly easily correctable flaws.

Unfortunately, this course of action appears to have been particularly prevalent in human geography. The dominant philosophy appears to be that most types of spatial modelling efforts are fatally flawed because they fail to account for the complex attitudes, preferences and tastes of individuals. These latter attributes are influenced not only by personal circumstances and characteristics, but also by the cultural, social and political milieu in which individuals make spatial decisions. A classic target of such criticism is the mathematical modelling of movements over space, more formally known as spatial interaction modelling or spatial choice modelling (Fotheringham and O'Kelly, 1989; Fotheringham 1991a; Sen and Smith, 1995). This is despite the fact that spatial interaction modelling is one of the most applied geographical techniques (see, for instance, the range of applications and references in Fotheringham and O'Kelly, 1989). It is used in studies of retail location, in impact assessment, in forecasting the demand of housing, in regional population projections, in forecasting travel demand within urban areas, and in a host of other applied areas. Consequently in the remainder of this chapter, we describe the evolution of spatial interaction models and their theoretical basis to demonstrate three issues:

1   That spatial modellers are aware of the need to make models more realistic in terms of human behaviour and spatial interaction models provide a classic example of this type of evolution.

2   The research frontier in spatial interaction modelling has progressed well beyond that perceived by those who still view spatial interaction models as 'gravity models' using outdated ideas from social physics. Spatial interaction/ spatial choice models provide a salutary lesson to those who would have us discard such approaches in human geography on how one type of spatial modelling can evolve in terms of its theoretical basis to provide fascinating insights into human spatial behaviour.

3   How the research frontier in this area has been dynamic with profound shifts in the theoretical bases used to justify the modelling approach. Spatial interaction modelling provides a good example of the evolution of quantitative geography from the stage of being fuelled primarily by the importation of ideas from other disciplines to a stage in which it is the seedbed for new spatial theories.

Four distinct phases of spatial interaction modelling can be identified, each of which has provided a quantum leap in our understanding of spatial movements over the previous era. In chronological order, these phases are: (a) spatial interaction as *social physics*; (b) spatial interaction as *statistical mechanics*; (c) spatial interaction as *aspatial information processing*; and (d) spatial interaction as *spatial information processing*. The approximate periods in which each of these theoretical bases dominated our understanding of spatial interaction are shown in Figure 9.1.

The major criticisms levelled at spatial interaction modelling stem from the first three phases when geography borrowed heavily from concepts developed in other disciplines. Indeed, there was a lingering suspicion, based on empirical evidence on the spatial pattern of local parameter estimates, that the basic formulation for spatial interaction models was a gross misspecification of reality. It is only with recent developments in understanding the basis of spatial interaction models from the perspective of spatial information processing that we have been able to identify and correct this misspecification. The identification of the misspecification bias in traditional spatial interaction models thus provides a strong link to the types of research described in Chapter 5 on local analysis. It was only through the calibration of *local* forms of spatial interaction models that the symptoms of model misspecification became evident: they were completely hidden in the output of the global models. We describe the nature of the evidence from local spatial interaction models in Section 9.5, but for a fuller discussion, the reader is referred to Fotheringham (1981; 1983b; 1984; 1991b).

9.2 Spatial interaction as social physics (1860–1970)

Attempts at understanding regularities in patterns of spatial flows began as early as the mid-nineteenth century with Carey's (1858) observations, followed by Ravenstein's (1885), that the movement of people between cities was analogous to the gravitational attraction between solid bodies. That is, greater numbers of migrants

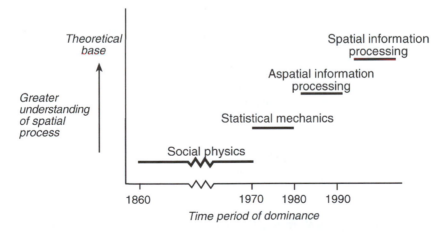

Figure 9.1   The changing theoretical basis of spatial interaction models

were observed to move between larger cities than smaller ones, *ceteris paribus*, and between cities which were closer together than being farther apart, *ceteris paribus*. This led to the proposal of a simple mathematical model to predict migration flows between origins and destinations based on the Newtonian gravity model. This model has the following form:

$$T_{ij} = k \frac{P_i P_j}{d_{ij}} \tag{9.1}$$

where T_{ij} represents the number of trips between origin i and destination j, P_i and P_j represent the sizes of origin i and destination j, respectively (as measured, for example, by their populations), d_{ij} represents the distance between i and j, and k is a scaling parameter relating the magnitude of T_{ij} to the ratio $P_i P_j / d_{ij}$. The units of k would be miles/person, for example, if distance were measured in miles and trips by the number of people going between i and j.

Subsequently, it was recognized that the relationships embedded in the simple spatial interaction model shown in (9.1) might vary according to the type of movement being investigated as well as to the economic and social milieu in which the movement was taking place. For example, the deterrence of distance in trip-making behaviour is likely to be greater in less developed economies with relatively poor transportation facilities than in more advanced economies with relatively good transportation. Similarly, individuals are generally more deterred by distance when shopping for basic goods such as bread and potatoes than when shopping for luxury items such as jewellery or antiques. Hence the basic formulation in (9.1) was modified to allow for these variations in behaviour:

$$T_{ij} = k \frac{P_i^{\alpha} P_j^{\lambda}}{d_{ij}^{\beta}} \tag{9.2}$$

where α, λ and β are parameters to be estimated empirically and reflect the nature of the relationship between spatial flows and each of the explanatory variables.

A further refinement of the basic gravity model formula resulted from the recognition that many origin and destination attributes, rather than simply the two size variables, are likely to influence flow patterns so that the model could be written as

$$T_{ij} = k \frac{V_{i1}^{\alpha 1} V_{i2}^{\alpha 2} \cdots V_{if}^{\alpha f} V_{j1}^{\lambda 1} V_{j2}^{\lambda 2} \cdots V_{jg}^{\lambda g}}{d_{ij}^{\beta}} \tag{9.3}$$

where there are f origin attributes, V_i, affecting the magnitude of the flows leaving i and g destination attributes, V_j, affecting the magnitude of flows entering j (see

Haynes and Fotheringham (1984) for more details). The models in (9.2) and (9.3) can easily be made linear by taking logarithms of both sides of the equations which allows calibration by OLS regression with the caveats described by Fotheringham and O'Kelly (1989).

Whichever formula, (9.1), (9.2) or (9.3), is used to analyse a set of spatial flows, the underlying model framework is basically the same and has its origins as a social science analogy to a physical model of gravitational attraction, the so-called 'social physics' approach to model building. As such, it has been justifiably criticized for its lack of any theoretical grounding in the way individuals behave. However, even though this derivation of the model is theoretically empty, the model itself produces reasonably accurate estimates of spatial flows. Therefore, a great deal of effort has been expended on trying to develop an acceptable theoretical framework for the gravity model.[1]

9.3 Spatial interaction as statistical mechanics (1970–80)

While various frameworks were espoused in attempts to justify the mathematical form of the gravity model (see for instance those of Dodd, 1950; Zipf, 1949; Huff, 1959; 1963; Nierdercorn and Bechdolt, 1969), the next major advance in providing a theoretical base came with the work of Wilson (1967; 1975). Wilson's pioneering work, from his background in physics, produced what has become known as a 'family of spatial interaction models', each member of which can be derived from the principles of statistical mechanics (for further details, see Fotheringham and O'Kelly, 1989, Chapter 2).

Wilson considered a flow matrix, denoting the numbers of individuals moving between each origin–destination pair to be a 'macrostate' of the system. This macrostate results from the combination of many individual 'microstates' where a microstate in this context is defined as a description of the individuals moving between origins and destination. For instance, a macrostate description might be that five individuals moved from A to B and two moved from A to C; a microstate description would name the individuals. Clearly, there are many different micro-states that could produce the same macrostate. If there are N individuals to be assigned to a set of categories and the number of individuals in category i is N_i, then the number of microstates associated with any given macrostate is

$$R = N! / \prod_i N_i! \tag{9.4}$$

For example, in the above situation where we assign five individuals from A to B and two individuals from A to C, there are $7!/(5! \times 2!) = 21$ possible ways of doing this. Different macrostates can have different numbers of microstates associated with them. For instance, if we assigned all seven individuals from A to

B and none from A to C, the number of microstates is only 1. If we assign six individuals from A to B and one from A to C, the number of microstates yielding this macrostate is 7, and if we assign four individuals from A to B and three from A to C, the number of microstates is 35. A full description of the possibilities is shown in Table 9.1.

Wilson formulated the problem of deriving a mathematical model of spatial flows in terms of choosing the particular macrostate which can be constructed from the largest number of microstates. This would be most likely to occur in the absence of any other information. In the above problem, this would be either four individuals going from A to B and three going from A to C or three going from A to B and four going from A to C. In a spatial flow context, the problem is then to select the set of T_{ij} values which maximizes R in Equation (9.4). Equivalently, one can maximize the natural logarithm of R divided by T, the total number of trips in the system (this will not alter the optimization result but is mathematically more tractable). This can be written as

Find the set of T_{ij} values which maximizes

$$H \equiv (1/T) \ln R = (1/T) \left(\ln T! - \sum_{ij} \ln T_{ij}! \right) \qquad (9.5)$$

where ln represents a natural logarithm. If all the T_{ij} values are large, use can be of Stirling's approximation that

$$\ln T! = T \ln T - T \qquad (9.6)$$

and

Table 9.1 **An example of microstates and macrostates: allocate seven individuals between two flows (A $\longrightarrow$ B; A $\longrightarrow$ C)**

Macrostate		Number of microstates
A $\longrightarrow$ B	A $\longrightarrow$ C	
7	0	1
6	1	7
5	2	21
4	3	35
3	4	35
2	5	21
1	6	7
0	7	1

$$\ln T_{ij}! = T_{ij} \ln T_{ij} - T_{ij} \qquad (9.7)$$

to obtain

$$H = (1/T)\left(T \ln T - T - \sum_{ij} T_{ij} \ln T_{ij} + T \right) \qquad (9.8)$$

On rearranging, this can be written as

$$H = -\sum_{ij}(T_{ij}/T)\ln(T_{ij}/T) \qquad (9.9)$$

or, equivalently, defining p_{ij}, the proportion of all trips that originate at i and terminate at j, as T_{ij}/T,

$$H = -\sum_{ij} p_{ij} \ln p_{ij} \qquad (9.10)$$

which is the formula for the entropy of a distribution (Shannon, 1948; Jaynes, 1957; Georgescu-Roegen, 1971) and which can be interpreted as a measure of the uncertainty about which microstate actually produces the observed macrostate.

However, it is perhaps obvious from the results in Table 9.1 that finding the set of trips which maximizes the entropy measure in Equation (9.9) or (9.10) is actually a trivial matter – H will be at a maximum when the T_{ij} values are all equal (or as near to equal as is possible). Wilson's further contribution to the theoretical derivation of spatial interaction models was to add constraints to this maximization procedure. The constraints that have been imposed on the system are

$$\sum_{ij} T_{ij}^* \ln P_i = P_1 \qquad (9.11)$$

where T_{ij}^* represents the model prediction of T_{ij} and P_i is the population of origin i;

$$\sum_{ij} T_{ij}^* \ln P_j = P_2 \qquad (9.12)$$

$$\sum_{ij} T_{ij}^* \ln d_{ij} = D \qquad (9.13)$$

where D is the total distance travelled in the system;

$$\sum_{ij} T_{ij}^* = K \tag{9.14}$$

where K is the known total interaction;

$$\sum_j T_{ij}^* = O_i \qquad \text{for all } i \tag{9.15}$$

where O_i is the known total flow from each origin; and

$$\sum_i T_{ij}^* = D_j \qquad \text{for all } j \tag{9.16}$$

where D_j is the known total inflow into each destination.

Imposing different combinations of these constraints on the maximization of Equation (9.9) produces different types of spatial interaction models – the so-called 'family' of spatial interaction models (Wilson, 1974; Fotheringham and O'Kelly, 1989). For instance, maximizing (9.9) subject to the constraints in (9.11) to (9.14) produces the gravity model formulation in Equation (9.2). Maximizing (9.9) subject to (9.12), (9.13) and (9.15) produces what is known as a production-constrained spatial interaction model:

$$T_{ij} = O_i P_j^\lambda d_{ij}^\beta \bigg/ \sum_j P_j^\lambda d_{ij}^\beta \tag{9.17}$$

Maximizing (9.9) subject to (9.11), (9.13) and (9.16) produces what is known as an attraction-constrained model:

$$T_{ij} = D_j P_i^\alpha d_{ij}^\beta \bigg/ \sum_i P_i^\alpha d_{ij}^\beta \tag{9.18}$$

and maximizing (9.9) subject to (9.13), (9.15) and (9.16) produces a production–attraction-constrained or doubly constrained model:

$$T_{ij} = A_i O_i B_j D_j d_{ij}^\beta \tag{9.19}$$

where

$$A_i = \sum_j (B_j D_j d_{ij}^{\beta})^{-1} \qquad (9.20)$$

and

$$B_j = \sum_i (A_i O_i d_{ij}^{\beta})^{-1} \qquad (9.21)$$

Examples of the applicaton of these models and details on how to calibrate them are given in Fotheringham and O'Kelly (1989). Using a framework first recognized by Alonso (1978), Fotheringham and Dignan (1984) show how each of these four spatial interaction models can be considered as the extremal points on a continuum of models defined in terms of how strictly the various constraint equations are enforced.

The importance of Wilson's entropy-maximizing derivation of a family of spatial interaction models was that it provided a theoretical justification for what had been until that time only an empirical observation. However, there are several criticisms which can be levelled at the derivation. First, it simply replaces one physical analogy – that of gravitational attraction – with another – that of statistical mechanics. The derivation is still sterile in terms of the processes by which individuals make spatial decisions. Secondly, while some of the constraint equations have a behavioural interpretation, others, such as those on the populations, are more difficult to justify except that they lead to a particular model form. Thirdly, the use of Stirling's approximation to derive the entropy formulation is highly suspect in most situations because of its assumption that all the T_{ij} values are large. When the T_{ij} values are small, the approximation is rather poor.

The latter criticism has been addressed within this framework by Webber (1975; 1977), Tribus (1969) and Dowson and Wragg (1973), amongst others, who eliminate the need to derive the entropy formulation from a discussion of microstates and macrostates. They argue that the entropy formulation satisfies certain reasonable requirements of a measure of statistical uncertainty and therefore they simply state this as a starting point for the derivation of mathematical models, the argument being that we should maximize our uncertainty about any outcome subject to constraints. To do otherwise would be adding bias into the model-building procedure. Despite this argument, however, there are still considerable difficulties with entropy maximization as a framework for the development of models of human spatial behaviour. Consequently, despite stimulating large amounts of research into spatial interaction modelling, the Wilson framework has been largely discarded in favour of more behavioural approaches.

9.4 Spatial interaction as aspatial information processing (1980–90)

The major advance in giving spatial interaction models a theoretical foundation that was based on human behaviour and information processing was the recognition that models such as those in (9.17) and (9.18) are 'share' models and have a logit formulation. The model in (9.17), for example, which is the most widespread form of spatial interaction model in use today, allocates shares of the number of people leaving an origin, O_i, amongst the destinations according to the attributes of these destinations. The derivation of this model could then be based on the derivation of the discrete choice model by McFadden (1974; 1978; 1980).

Consider an individual at origin i about to make a choice of a single destination j from a set of possible destinations. Suppose the individual evaluates every alternative and then selects the one yielding him or her maximum benefit or *utility*. The utility accruing from a person at i selecting alternative j, U_{ij}, can be thought of as being composed of two parts, a measurable component, V_{ij}, and an unmeasurable component, μ_{ij}. That is,

$$U_{ij} = V_{ij} + \mu_{ij} \qquad (9.22)$$

If we knew the values of U_{ij} for every alternative we could say with certainty which destination an individual would choose. That is,

$$p_{ik} = \begin{cases} 1 \text{ if } U_{ik} > U_{ij} \text{ (for all } j \in N, j \neq k) \\ 0 \text{ otherwise} \end{cases} \qquad (9.23)$$

where k is a specific alternative from the set of N alternatives. However, given that each alternative has an unknown component to its utility, we cannot say for certain which alternative an individual will choose. We can only evaluate each alternative based on the measurable component and then state a probability that a destination will be selected based on a comparison of the observable components of the utility functions. That is,

$$p_{ik} = \text{prob}[U_{ik} > U_{ij} \text{ (for all } j \in N, j \neq k)] \qquad (9.24)$$

which, on substituting (9.22) and rearranging, can be written as

$$p_{ik} = \text{prob}[\mu_{ij} < V_{ik} - V_{ij} + \mu_{ik} \text{ (for all } j \in N, j \neq k)] \qquad (9.25)$$

Recognizing that the μ_{ik} and the μ_{ij} terms are random elements drawn from a continuous distribution ranging from $-\infty$ to $+\infty$, Equation (9.25) can be written as

$$p_{ik} = \int_{x=-\infty}^{+\infty} g(\mu_{ik} = x) \prod_{j(j \neq k)}^{n} \int_{y=-\infty}^{V_{ik} - V_{ij} + x} g(\mu_{ij} = y)\, \mathrm{d}y\, \mathrm{d}x \qquad (9.26)$$

where $g(\ )$ represents some probability density function. McFadden (1974) demonstrated that if the μ terms are distributed according to a Type I extreme value distribution (Fisher and Tippett, 1928), the model which results is

$$p_{ik} = \exp(V_{ik}) / \sum_{j} \exp(V_{ij}) \qquad (9.27)$$

and which has a behaviourally intuitive logistic response rate between the probability of choosing a particular alternative and its observed utility (see Fotheringham and O'Kelly (1989, p. 74) for a demonstration of this).

The expression in (9.27) has the same basic form as the spatial interaction models in (9.17) and (9.18). To see this more clearly, the relationship can be made more explicit by defining the observable utility associated with each destination as a function of its attributes. For example, consider the case of retail choice in which an individual selects a supermarket from a set of supermarkets in an urban area. The individual is likely to gain more benefit from selecting a large store as opposed to a small store in terms of having a greater variety of produce to choose from and also possibly cheaper prices. Similarly, the individual will gain more benefit, in terms of cost and time savings, from selecting a store in close proximity rather than far away. Obviously, there are many other attributes that can be added to an individual's utility function but these two will serve our purpose in demonstrating the behavioural relationships embedded in the model derivation at this stage.

Consider first the relationship between V_{ik} and S_k, the size of the store. Two possibilities are shown in Figure 9.2: a linear relationship and a logarithmic relationship. The linear relationship might at first seem the more reasonable but on reflection the behaviour it represents is rather implausible. It implies that an additional 1000 square metres of retail space to an existing store would generate the same added utility regardless of the original size of the store. That is, the additional utility from the extension is predicated to be the same whether the original store is 2000 square metres or 10 000 square metres. An alternative and more plausible behavioural relationship is that provided by the logarithmic function which implies that the addition to the size of the store generates less benefit to the consumer when the store is already very large than when the store is small. Consumers are unlikely to reap much benefit from an addition when the store is already very large. Hence, the relationship between V_{ik} and S_k can be expressed as

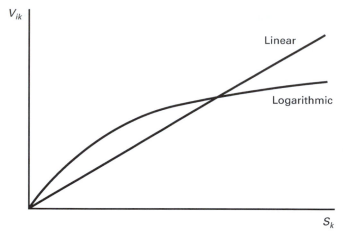

Figure 9.2   **The relationship between observed utility and size**

$$V_{ik} \propto \alpha \ln S_k \qquad (9.28)$$

where α represents the particular shape of the logarithmic relationship.

For a similar behavioural reason, the relationship between V_{ik} and d_{ik}, the distance to alternative k for an individual at i, is also likely to be logarithmic as depicted in Figure 9.3. The addition of 10 kilometres to a journey which is already 500 kilometres is likely to affect one's assessment of the journey less than the addition of 10 kilometres to a journey which is only 5 kilometres. Hence, the appropriate relationship is given by

$$V_{ik} \propto \beta \ln d_{ik} \qquad (9.29)$$

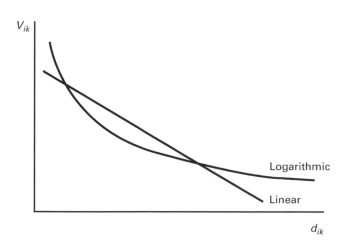

Figure 9.3   **The relationship between observed utility and distance**

where β is a distance-decay parameter to be estimated. Substituting the relationships described by Equations (9.28) and (9.29) into (9.27) yields

$$p_{ik} = S_k^\alpha d_{ik}^\beta \Big/ \sum_j S_j^\alpha d_{ij}^\beta \qquad (9.30)$$

which is the form of production-constrained spatial interaction model given in (9.17).

It should be noted that this form of model has two well-known properties which, in the context of spatial choice, are undesirable. The first is termed the independence from irrelevant alternatives (IIA) property and stated simply is that the ratio of the probabilities of an individual selecting two alternatives is unaffected by the addition of a third alternative. This is easy to see: in Equation (9.30) consider two alternatives, j and k. As the denominator in (9.30) is a constant, the ratio of the probabilities of choosing k over j is

$$p_{ik} / p_{ij} = S_k^\alpha d_{ik}^\beta / S_j^\alpha d_{ij}^\beta \qquad (9.31)$$

which is indeed independent of any other alternative. In a spatial choice context this property implies that the model's predicted flow to any destination is unaffected by the location of the other alternatives. That is, the ratio of the probabilities of choosing two stores is unaffected by the fact that one store might be surrounded by competitors and the other store is in relative isolation. This seems unrealistic in terms of most spatial choice contexts where the location of an alternative with regard to its competitors would seem to be important (Fotheringham, 1989).

The other undesirable property of the model in (9.30) is what Huber et al. (1982) term 'regularity'. That is, it is impossible for the model to predict an increase in the flow to any existing alternative due to the addition of a new alternative. Again, in many spatial choice situations, this is unnecessarily restrictive and unrealistic. Consider, for example, clothing stores where agglomeration effects are present (one of the reasons why shopping malls exist). The location of a new store can increase the sales at nearby stores.

More recent developments in the theoretical derivation of spatial interaction models have used *spatial* choice principles, which avoid both of the above problems. These developments and the resulting models are now discussed.

9.5 Spatial interaction as spatial information processing (1990 onwards)

The derivation of spatial interaction models described above in the framework of discrete choice represents an advance over the previous physical analogies to gravitational attraction and entropy. However, it is a framework which has also been 'borrowed' from economics, another predominantly aspatial discipline. The discrete choice framework was developed for aspatial contexts such as brand choice and choice of transportation mode. As such, it contains an assumption that, while tenable in the context of most aspatial choice, is untenable when applied to most spatial choice problems. In Equation (9.24), and subsequently, the assumption is made that an individual is able to evaluate all alternatives. That is, alternative k is compared with all the alternatives in the full choice set N. In essence, the individual is assumed to be omniscient and able to process vast amounts of information. In simple choice situations where N, the number of alternatives, is small, such as in most brand choice and mode choice situations, it is reasonable to assume that individuals can process information on all the alternatives. However, it is generally recognized that individuals have a limited capacity for processing information (*inter alia*, Simon, 1969; Lindsay and Norman, 1972; Newell and Simon, 1972; Norman and Bobrow, 1975) and indeed this limit might be reached very quickly. Bettman (1979), for example, argues that our limit to information processing might be reached with as few as six or seven alternatives. Consequently, in most spatial choice situations the number of alternatives is far too large to assume individuals can evaluate all possible alternatives. Consider, for example, buying a place to live in a large metropolitan area where there might be thousands of properties for sale, or a migrant choosing between cities within a country, or even a consumer selecting a store within an urban area. In such situations the number of alternatives is generally very large, far too large in fact for us to be able to retain the assumption that individuals evaluate every alternative.

The manifestation of this untenable assumption and the subsequent misspecification of the traditional spatial interaction model formulation only comes to light when local forms of spatial interaction model are calibrated (see Chapter 5). Typically, local spatial interaction models are calibrated separately for each origin in the system (although they can also be calibrated separately for each destination). In so doing, a set of parameters, one for each origin, describes the relationship between spatial movement and an attribute of the destinations. These sets of origin-specific parameter estimates replace the single parameter estimates obtained in the calibration of a global model. An indication of a problem with the classical spatial interaction model formulation was uncovered when origin-specific estimates of the distance-decay parameter were mapped. The maps showed a consistent and puzzling spatial pattern in which peripheral origins generally had more negative distance-decay parameter estimates than more central ones (see Fotheringham (1981) for examples of this pattern and a summary of the discussions, sometimes intense, that such patterns evoked).

An example of the trend in origin-specific distance-decay parameter estimates commonly found is shown in Figure 9.4. This describes the relationship between origin-specific distance-decay parameter estimates and a measure of the centrality of each origin. The parameters are derived from a production-constrained model (such as that shown in Equation (9.17)) applied to 1970–80 migration flows between the 48 contiguous states of the USA. The destination attributes in the

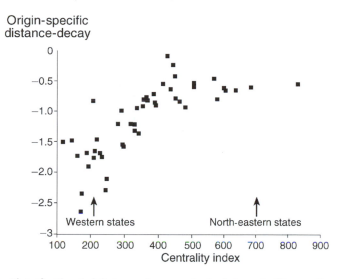

Figure 9.4   Plot of estimated distance-decay against origin centrality: results contain a misspecification bias

model are: population size; distance from each origin; mean personal income; average temperature; unemployment rate; and median house price. The pattern is quite striking in that it appears to indicate that distance-decay rates are less negative for centrally located states than for peripherally located states.[2] Behavioural explanations for the types of pattern exhibited in Figure 9.4, such as people in more central origins being more spatially mobile, have not been convincing (see Fotheringham, 1981) and the search turned to a theoretical explanation to this puzzle. The answer came in understanding the type of information processing which is likely to take place in *spatial* choice, which we now describe.

The generic spatial choice problem can be stated as: 'How does an individual at location *i* make a selection from a set of *N* spatial alternatives?' This type of spatial choice must precede most types of spatial interaction, whether it be shopping, migration, house selection which determines commuting patterns, vacationing, or any other of a myriad types of movement over space. The spatial choice process as shown in Figure 9.5 has three characteristics (Haynes and Fotheringham, 1990):

1   It is a discrete, rather than a continuous, process. That is, either a destination is selected or it is not and there is a finite number of alternatives.
2   The number of alternatives is generally large, and in some cases, very large.
3   The alternatives have fixed spatial locations, which limits the degree to which alternatives are substitutes for one another. It also means that, unless the spatial distribution of alternatives is perfectly regular, each alternative faces a unique spatial distribution of competing alternatives.

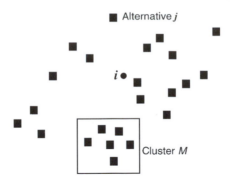

How does an individual at *i* select an alternative
from the set of *N* alternatives?

Figure 9.5 **The spatial choice problem**

The first property suggests that the discrete choice framework introduced from
aspatial choice, and described above in Section 9.4, provides a basis for understanding spatial choice. The second and third properties suggest the framework will
need some modification as these two properties are not usually found in aspatial
choice.

With the spatial choice framework in mind, and the realization that there are
limits to individuals' abilities to process information, Fotheringham (1983a; 1983b;
1984; 1986; 1991b) develops a new form of spatial interaction model, termed a
competing destinations model. This model results from the starting point of how
individuals process *spatial* information and its derivation does not depend on the
assumption that individuals are able to evaluate every alternative.

The derivation of the competing destinations model closely follows that of the
logit discrete choice model described above with the added flexibility of allowing
individuals to make choices from restricted choice sets rather than from the
complete set of alternatives. This is achieved by replacing Equation (9.24) with

$$p_{ik} = \text{prob}[U_{ik} > U_{ij} + \ln p_i(j \in M) \text{ (for all } j \in N, j \neq k)]p_i(k \in M) \quad (9.32)$$

where M is the restricted choice set in which an individual at i evaluates all the
alternatives and N is the full set of alternatives. Alternatives that are not within M
are not evaluated and therefore cannot be chosen by the individual. For example,
suppose an individual looking for a house within a city only looks at housing in the
south-east quadrant of the city. Houses in other parts of the city are never considered and so are not evaluated and therefore will not be chosen, even if they have all
the attributes the individual seeks. To see how Equation (9.32) represents this type
of behaviour, consider two situations. First, if the alternative k is not in the
restricted choice set M, $p_i(k \in M)$ will be zero and hence $p_{ik} = 0$, no matter how

large U_{ik} is. Secondly, if j is not in the restricted choice set, $p_i(j \in M) = 0$ and $\ln p_i(j \in M) = -\infty$ so that $U_{ij} + \ln p_i(j \in M) = -\infty$ and hence k will be selected in preference to j despite j's attributes. Thus, Equation (9.31) allows sub-optimal choices to be made: individuals sometimes select alternatives which do not yield maximum utility because they cannot evaluate every alternative.

Substituting Equation (9.32) for (9.24) into the framework described in Section 9.4 yields the following general spatial choice model:

$$p_{ik} = p_i(k \in M) \int_{x=-\infty}^{+\infty} g(\mu_{ik} = x) \prod_{j(j \neq k)}^{n} \int_{y=-\infty}^{V_{ik}-V_{ij}+x-\ln p_i(j \in M)} g(\mu_{ij} = y) \, dy \, dx \quad (9.33)$$

(Fotheringham, 1988). Under the assumption that the μ terms are independently and identically distributed with a Type I extreme value distribution, the resulting spatial choice model is

$$p_{ik} = \exp(V_{ik}) \times p_i(k \in M) \Big/ \sum_j \exp(V_{ij}) \times p_i(j \in M) \quad (9.34)$$

Essentially this is a logit model where each alternative's observable utility is weighted by the probability of the alternative being evaluated. Three specific models can be derived from this general form depending on the definition of $p_i(j \in M)$:

1 If $p_i(j \in M) = 1$ for all j, then Equation (9.34) is equivalent to (9.27). That is, if all the alternatives *are* evaluated by individuals, then the model in (9.34) degenerates to the classic logit model.

2 If $p_i(j \in M) = 1$ for all $j \in M$ and 0 otherwise, the model is equivalent to a logit model applied to only those alternatives in the restricted choice set. Obviously to operationalize such a model it would be necessary to know in advance what the restricted choice set was for each individual. Generally such information is not known a priori. Also, the situation is complicated by the likelihood that individuals in different locations have different mental images of space and different cognitive constructions of clusters of spatial alternatives (Gould and White, 1974).

3 If $p_i(j \in M)$ is a function of the location of j with respect to the other alternatives, then the model in (9.34) holds and the task is to determine how to define $p_i(j \in M)$. We now pursue this line of enquiry in order to derive the competing destinations spatial interaction model (Fotheringham, 1983b; 1986).

By defining a likelihood as a probability divided by a constant, the model in (9.34) can be rewritten as

$$p_{ik} = \exp(V_{ik}) \times l_i(k \in M) \bigg/ \sum_j \exp(V_{ij}) \times l_i(j \in M) \qquad (9.35)$$

where $l_i(j \in M)$ represents the likelihood that alternative j is in the restricted choice of alternatives which are evaluated by an individual at location i. A great deal of research has been undertaken into how to define this likelihood function, or its equivalent (*inter alia*, Meyer, 1979; Batsell, 1981; Meyer and Eagle, 1982; Fotheringham, 1983b; 1991b; Fik and Mulligan, 1990; Lo, 1991; Fik et al., 1992; Thill, 1992).

It can be argued that there are two distinct processes represented in Equation (9.35): one is *how* individuals evaluate an alternative and the other is *whether* it is evaluated (see Figure 9.6). Associated with these two processes are three types of attributes:

1   The attributes of an alternative that affect *how* individuals evaluate it. These are already incorporated in the model through the expression $\exp(V_{ij})$, the exponential of the observable utility component.
2   The attributes of an alternative that affect both *how* individuals evaluate it and *whether* they evaluate it. Again, such attributes are already included in the spatial choice framework through the expression $\exp(V_{ij})$.
3   The attributes of an alternative that affect only *whether* an alternative is evaluated. It is these attributes which need to be incorporated into the

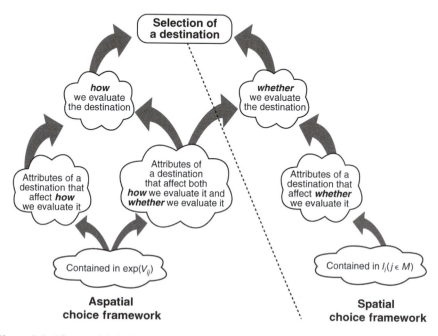

Figure 9.6   **The spatial choice process**

modelling framework through the definition of the likelihood of an alternative being evaluated, $l_i(j \in M)$. The definition of these attributes will lead to new and improved forms of spatial choice/spatial interaction models. We now describe the rationale behind the definition of one of these attributes which describes the location of an alternative *vis-à-vis* all other alternatives and which leads to a model known as the competing destinations model (Fotheringham, 1983b; 1986).

Assume that, because of limits on our ability to process information, individuals do not evaluate all the alternatives available in a typical spatial choice context. Instead, they process spatial information and make spatial choices in such a way that they first evaluate clusters of alternatives and only evaluate particular alternatives from within a selected cluster. For instance, a migrant might first select a region in which he or she would like to live and then evaluate alternatives only within that region. Equivalently, an individual looking at housing in a city might have strong opinions about which parts of the city he or she would like to live in and which parts are to be avoided. Support for the hypothesis of hierarchical information processing is given by, *inter alia*, McNamara (1986; 1992), Walker and Calzonetti (1989), Hirtle and Jonides (1985), Stevens and Coupe (1978), Curtis and Fotheringham (1995) and Fotheringham and Curtis (1992; 1999).

Unfortunately for the modeller it is very difficult, if not impossible, to know how people mentally cluster alternatives in space. Individuals themselves might not be able to tell you how they process spatial information and the mental clusters they form might well be fuzzy (Zadeh, 1965; Pipkin, 1978) rather than discrete. Certainly, there is evidence to suggest that such clusters will vary according to one's location in space (Gould and White, 1974; Downs and Stea, 1977; Curtis and Fotheringham, 1995; Fotheringham and Curtis, 1999). However, Fotheringham (1991b) and Pellegrini and Fotheringham (1999) argue that it is not necessary to define the clusters of alternatives that individuals evaluate. It is only necessary to assume that some form of hierarchical spatial information processing takes place whereby clusters of alternatives are first evaluated and then only alternatives within a selected cluster are evaluated. The task then becomes one of identifying any destination attributes associated with the hierarchical processing of spatial information that affect the likelihood of an alternative being evaluated.

One such variable, identified by Fotheringham (1983b; 1986), is the proximity of an alternative to all other possible alternatives. There are many ways such a variable could be measured, but a common measurement is the accessibility of an alternative k to all other alternatives, A_k, as defined by

$$A_k = \sum_{j \neq k} P_j^\alpha / d_{jk}^\beta \qquad (9.36)$$

where P_j represents the population of alternative j and d_{jk} is the distance between alternatives j and k. This formulation measures the 'competition' faced by alternative k from all the other alternatives: when an alternative is centrally located near many other alternatives, A_k will be large; when it is relatively isolated, A_k will be small. In practice, the values of α and β are often set to 1 and -1, respectively, although they could be estimated either in the model calibration procedure or externally.

The justification for the inclusion of this competition variable is based on 'the psychophysical law' that individuals tend mentally to underestimate the size of large objects (Stevens, 1957; 1975). Consider an individual evaluating spatial clusters of alternatives, such as cities within a country. One of the factors in the evaluation of each cluster is likely to be its size, reflecting the number of opportunities for employment etc. However, the psychophysical law suggests that the cognized size of a large cluster is likely to be smaller than its objective size and that the magnitude of this mental underrepresentation will increase with the size of cluster, resulting in the logarithmic relationship depicted in Figure 9.7. The linear relationship between the cognized and objective size of clusters would only occur if individuals did not underrepresent large clusters mentally. The probability of an individual selecting a large cluster of alternatives, and consequently evaluating the individual alternatives within that cluster, will obviously be less if the relationship is logarithmic than if it were linear. Hence, the likelihood of a spatial alternative being evaluated is a function of the location of an alternative with respect to all the other alternatives; alternatives in large clusters are less likely to be evaluated than those in small clusters, *ceteris paribus*. Rather than having to identify the exact nature of the clusters cognized by individuals, all we therefore need to define $l_i(j \in M)$ is the measure of destination competition given in Equation (9.36). That is,

$$l_i(j \in M) = A_j^\delta \qquad (9.37)$$

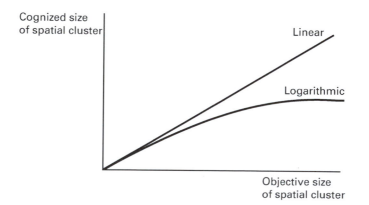

Figure 9.7   **Relationships between cognized and objective cluster sizes**

where δ is a parameter to be estimated. If $\delta = 0$, all destinations are equally likely to be evaluated (the aspatial logit assumption). Given the above theoretical reasoning, we would expect that δ will be negative in most spatial choice situations; the more centrally located is an alternative, the more likely it is to be in a large cluster cognized by an individual and the less likely it is to be evaluated, *ceteris paribus*. However, for certain types of spatial choice, such as non-grocery shopping, there could well be substantive agglomeration effects so that individuals are attracted to large clusters of alternatives in order to minimize the costs of comparison shopping. In such instances, δ will be positive.

Substituting (9.37) into (9.35) yields the spatial choice model,

$$p_{ik} = \exp(V_{ik}) \times A_k^{\delta} \bigg/ \sum_j \exp(V_{ij}) \times A_j^{\delta} \tag{9.38}$$

which Fotheringham (1983; 1986) calls the 'competing destinations model'. A large number of empirical studies have demonstrated the superiority of this model over the aspatial logit formulation in modelling spatial choice (*inter alia*, Fotheringham 1984; 1986; Fotheringham and O'Kelly, 1989; Boyle et al., 1998; Atkins and Fotheringham, 1999; Pellegrini and Fotheringham, 1999). Typically, it produces demonstrably less biased parameter estimates and significant improvements in goodness of fit.

As an example, consider the pattern of origin-specific distance-decay parameters described above in Figure 9.4. This pattern was described as evidence of a misspecification problem with the model because there appears to be a regular change over space in the effect of distance on spatial interaction which is difficult to account for in behavioural terms. When a model having the form of (9.38) is calibrated with exactly the same destination attributes as those used to generate the results in Figure 9.4, the spatial pattern of the origin-specific distance-decay parameters disappears as shown in Figure 9.8, drawn to the same scale. This suggests strongly that the distance-decay parameters generated by standard logit forms of spatial interaction models contain a severe misspecification bias and that this bias is eliminated in the calibration of the competing destinations model. Fotheringham (1986) demonstrates theoretically the nature of the misspecification bias in the parameter estimates obtained from the aspatial logit model when the relationship in (9.38) holds.

By incorporating the appropriate spatial weight on each observable utility function, both of the undesirable properties identified with the aspatial logit model in the previous section, the IIA property and regularity, are also removed in the competing destinations model. The removal of the former can be seen by taking the ratio of the probabilities of selecting two alternatives, j and k, from the model in (9.38), which is

$$p_{ik}/p_{ij} = \exp(V_{ik}) \times A_k^{\delta}/\exp(V_{ij}) \times A_j^{\delta} \tag{9.39}$$

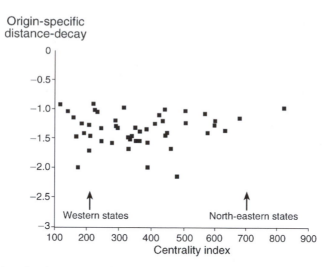

Figure 9.8 **Plot of estimated distance-decay against origin centrality: results without a misspecification bias**

and which is unlikely to be constant under the addition of a new alternative to the system. The addition of a new spatial alternative will have differential effects on the measurement of A_k and A_j depending on the location of the new alternative *vis-à-vis* the locations of k and j. Similarly, the regularity property is removed because within the competing destinations framework, it is possible that the addition of a new spatial alternative will result in an increase in the probability of selecting an existing alternative. This would happen, for instance, in a shopping context in which δ were strongly positive; then, the addition of a new store in close proximity to an existing one could result in an increase in the predicted patronage of the existing store.

9.6 Summary

The development of spatial interaction models highlights the progress made over the last few decades in this area of quantitative geography. We have witnessed the progression from models whose only justification was an empirical regularity and an analogy to gravitational attraction; to models derived from either non-behavioural or aspatial theories imported from other disciplines; and finally to models based on principles of spatial information processing, sub-optimality, hierarchical decision making and spatial cognition. Current research lies in consolidating and improving this latest framework.

However, despite the progress in making spatial interaction models more behaviourally based, it is probably the case that many geographers still associate spatial interaction modelling with its early social physics background (as demonstrated by the lingering usage of the term 'gravity model'). This is unfortunate for two reasons. The first is that, despite its very widespread application to many facets of the real world, these geographers ignore or even dismiss spatial interaction modelling, not because of what it is but because of what it was 20 or 30 years ago. The second, and more important, is that spatial interaction modelling provides a very fertile area for understanding spatial behaviour and for developing theories which are explicitly spatial. It is an area that is quintessentially geographical; it is an area where geographers should be leading the way by exporting their ideas to other disciplines.

Notes

1. Additionally, efforts in obtaining data for trip models have led to other spinoffs. For example, in the UK the first file linking postcodes to grid references arose out of government-sponsored trip modelling efforts (Raper et al., 1992, p. 43).
2. Centrality here is measured with respect to overall population so that in this US example, the states with the highest centrality index are those in the north-east and the states with the lowest centrality index are those in the west.

10 Challenges in Spatial Data Analysis

10.1 Retrospect

In 1963 Burton described the transformation of the 'spirit and purpose' of geography during the 1950s as 'the "quantitative revolution"', and commented that by the time he was writing 'the revolution itself is now over'. Later in the same paper Burton comments: 'An intellectual revolution is over when accepted ideas have been overthrown or have been modified to include new ideas'. However, others have identified the 'end' of the revolution as the starting point for moving away from quantification. This final chapter begins by examining the concerns of spatial analysts in the years following the 'revolution'. We shall consider whether or not those problems which were of major concern have been set aside as not worth expending effort on, whether they have been 'solved' with the solutions in common use, or whether they remain problems.

If the methodologies and techniques of the quantitative revolution were in place by Burton's 1963 paper, then a glance at some of the texts emerging in the late 1960s gives a flavour of matters of concern that had arisen even at this early stage. For example, King (1969) provides a helpful summary of 'statistical analysis' as practised in geography by the late 1960s. We find a conglomerate of simple descriptive statistics, a consideration of theoretical distributions, point pattern analysis using nearest-neighbour and quadrat methods, correlation, regression, residual mapping, trend surface analysis, principal components analysis, factor analysis, classification, and discriminant analysis. However, recognizing that this battery of statistical methods might still not be sufficient, King draws attention to a number of problems: (a) modifiable areal units, (b) spatial autocorrelation, (c) model identification, and (d) the use of dummy variables to represent regional effects. Thirty years on, these remain areas that pose some interesting challenges. Although we have made considerable progress in some of these areas, and we now know a great deal more about spatial data and spatial data analysis, this text has exposed how complex some of these issues are. We now briefly review some of the topics in which we see major challenges and opportunities in spatial analysis.

10.2 Current challenges

10.2.1 The modifiable areal unit problem

One problem that has long been identified in the analysis of spatially aggregated data is that for some analyses the results depend on the definition of the areal units for which data are reported (Gehlke and Biehl, 1934). This has become known as the modifiable areal unit problem (Openshaw, 1984). The data sets which are available for the solution of particular problems have often been aggregated to some system of zones by the agency responsible for collecting them. Notable examples here are the Census of Population data available in both the USA and the UK (although samples of microdata are available, they have limited applicability, particularly to spatial problems). Gehlke and Biehl (1934) noted that aggregation from census tracts to larger areal units increased the value of a correlation coefficient, and posed the question as to whether the correlation coefficient was an appropriate statistic for use with aggregate spatial data. Yule and Kendall (1950, pp. 310–314) observed the same phenomenon in an analysis of wheat and potato yields among the counties of England. In the same year Robinson (1950) discovered similar behaviour in comparing correlations based on individual measurements and those aggregated to counties in states. Robinson (1956) suggested that areal weighting would ameliorate the problems in calculating correlation coefficients for areal units of different sizes. Blalock (1964) observed that regrouping county data for 150 southern US counties into larger areal units increased the correlation coefficient. However, Thomas and Anderson (1965) re-examined the findings of Gehlke and Biehl, and Yule and Kendall, concluding that the differences in the values of the correlation coefficients in both studies were due to random variation.

Openshaw and Taylor (1979) discovered that they could obtain almost any value of the correlation between voting behaviour and age in Iowa merely by aggregating counties in different ways. More recently, Fotheringham and Wong (1991) demonstrated how the stability or instability of parameter estimates obtained from a multivariate model under different levels of aggregation could be visualized. They demonstrated that some relationships can be relatively stable to data aggregation, while others appear to be very sensitive. Fotheringham et al. (1995) describe the same problem in the context of location–allocation modelling where sensitive locational decisions are shown to be affected by the definition of the zones for which data are reported. Both these latter two studies also demonstrate the two components of the modifiable areal unit problem:

1 *The scale effect*: different results can be obtained from the same statistical analysis at different levels of spatial resolution.
2 *The zoning effect*: different results can be obtained owing to the regrouping of zones at a given scale.

However, despite the fact that the modifiable areal unit problem has been observed for decades, we appear to be little closer in dealing with the problem effectively. Of course, one solution is to use spatially disaggregated data, although, as mentioned above, such data are rarely available. Another is to report the results of the spatial analysis at the most spatially disaggregated level possible and then to demonstrate visually the sensitivity of the results to both the scale and zoning effects. In this way, if some results can be shown to be relatively stable over a wide range of zoning systems, this can induce greater confidence that the results at the most disaggregated level have some meaning and are not simply artefacts of the way the data are arranged.

For Openshaw and Rao (1995), the answer lies in the construction of zoning systems that are in some sense 'optimal'. It is not always clear whether this construction is intended to influence those releasing the data. It might be regarded as somewhat eccentric to create a zoning system for which the fit of a particular spatial model is 'optimal' if this zoning system is to be used for some completely different purpose. Equally, if the reaggregation is to take place after the data are released we know that we are stuck with the areal units that we have, and that if we reaggregate them to some other units, we are essentially creating arbitrary results. If we create one zoning system for one regression model, then include another variable and rezone to get the 'best fit', then it is not immediately clear how we compare one model with another, apart from in the rather trivial sense that the r^2 has changed.

One methodology that shows some promise is by Steel and Holt (1996), Holt et al. (1996) and Tranmer and Steel (1998). They propose a model structure to deal with the problems of modifiable areal units in which there is an extra set of *grouping variables* included. These grouping variables, z, can be measured at the individual level and are in some way related to the processes being measured at the aggregate level. They are used to adjust the aggregate-level variance–covariance matrix for the model so that it more closely approximates the unknown, individual-level variance–covariance matrix. Data on each of the following are assumed to be available:

1   A population of N individuals, $i = 1 \ldots N$, distributed over M regions, $j = 1 \ldots M$.
2   A vector of variables of interest for each individual, y_i, which has been aggregated to M zones. The aggregated data are represented as y_j.
3   A matrix **x** indicating to which of the regions each individual belongs. This matrix has typical element x_{ij}.
4   The number of individuals in the jth region, N_j.
5   A vector of variables z_i measured at the level of the individuals and defined as the grouping variables.

At the aggregate level, the variance–covariance matrix of the variables is $\mathbf{S}_{yy}$ and it is this that is of interest, since from it we can estimate a correlation matrix, regression coefficients and principal components of **y**. However, if we can estimate an individual-level variance–covariance matrix for the grouping variables, a better estimator of the aggregate-level variance–covariance matrix is given by

$$\Sigma_{yy} = \mathbf{S}_{yy} + \mathbf{B}_{yz}^T(\Sigma_{zz} - \mathbf{S}_{zz})\mathbf{B}_{yz} \qquad (10.1)$$

where Σ_{yy} is the new estimate of the aggregate-level variance–covariance matrix of the y variables; $\mathbf{S}_{yy}$ is the original estimate of the aggregate-level variance–covariance matrix of the y variables from the aggregated data; $\mathbf{B}_{yz}$ is a matrix of aggregate-level regression coefficients of y on z, the grouping variable; Σ_{zz} is the individual-level variance–covariance matrix for the z variables; and $\mathbf{S}_{zz}$ is the aggregate-level variance–covariance matrix for the z variables.

The problem, of course, is to find an estimate of the individual-level covariance matrix for suitable z variables. In some circumstances it may be possible to obtain individual-level data; Holt et al. (1996), for example, use data from the 1991 UK Census of Population for 371 enumeration districts in the administrative district of Reigate, Banstead and Tandridge (in South London) and they make use of the 'Sample of Anonymised Records' (Dale and Marsh, 1993) to provide individual-level records for a 2% sample of the district's population to estimate the Σ_{zz} matrix. The technique could be applied to US census data using the PUMS to estimate the individual-level variance–covariance matrix. The technique obviously does depend on the availability of individual-level data although Holt et al. (1996) point out that these need not be for the same area. If suitable individual-level data are not available, the technique is of little use. It should also be noted that the measurement of both the Σ_{zz} and the $\mathbf{B}_{yz}$ matrices depends on some definition of the areal extent of the individual-level variables and must therefore also be prone to the modifiable areal unit problem. It is also assumed that the relationships between the grouping variables and the aggregate variables are constant over space.

A not too dissimilar approach has been taken by King (1997) who provides one interesting 'solution' to the ecological fallacy problem in a voting context. Here, data are available at different spatial scales (voting precincts and districts) and King's solution involves local forms of regression (see Section 10.2.2) and using constraints peculiar to voting data. The use of local forms of analysis in general might well provide a new way of thinking about the modifiable areal unit problem. Geographically weighted regression, for example, provides fuzzy zones around each calibration point and presents a surface of parameters which is independent of any predefined zoning system. Of course, GWR still suffers from its own form of scale problem in that the results are dependent on the bandwidth used which is a measure of spatial scale. However, at least this can be calibrated within a regression framework.

Despite the promise shown by the above studies, and also possibly by the use of local forms of analysis in general, there is still much work to be done in exploring the sensitivity of the results of spatial analysis to the definition of spatial units for which data are defined. Until a general solution is found, the modifiable areal unit problem will continue to create uncertainty in the widespread applicability of spatial analytical results.

10.2.2 Spatial non-stationarity

The exploration of spatial variations in relationships has perhaps been one of the most fruitful areas of research in quantitative geography over recent years. The spatial weighting framework of GWR, in particular, seems to hold much promise (see Chapter 5). Although currently utilized as a tool for investigating spatial non-stationarity within a regression framework, the technique can be applied quite easily to other types of analysis.

For instance, the calculation of a simple mean produces a single whole-map statistic which summarizes the central tendency of a distribution, but which ignores any spatial effects. Suppose we allow the mean to vary across the map, and we explore the data by examining local variation in the mean. One possibility for a locally weighted mean might be something of the form

$$\tilde{x}(z) = \frac{\sum_{i=1}^{n} w_i(z) x_i}{\sum_{i=1}^{n} w_i(z)} \tag{10.2}$$

where $\tilde{x}(z)$ is the locally weighted mean calculated at the location whose coordinates are given by the vector $\mathbf{z}$. The $w_i(z)$ are weights for each observation calculated in the manner used in geographically weighted regression (see Chapter 5). Do local means have an intuitive interpretation, and what are the problems in using such statistics? To take housing cost as an example, some areas are colloquially described as 'expensive' compared with other areas. The houses in an 'expensive' area might be so because they are larger and have more amenities than other houses, but equally, there may well be a greater local demand for such properties thus increasing the unit cost of floorspace. A map showing the variation in the local mean price might provide some insight here that aggregating the data to a set of zones and calculating a mean or median might hide. One problem with a spatial mean is that of choosing a parameter (the bandwidth) for the spatial weighting function. The calibration methods of Chapter 5 based on cross-validation are not appropriate here as it is unclear what objective function is to be minimized. However, a map with an associated slider bar allowing a range of bandwidths to be chosen by the analyst might provide some insight into the spatial scale at which an area is considered 'expensive' or otherwise. Instead of imposing an arbitrary zoning system onto the data for analysis, we allow the data to suggest one instead.

No analysis involving a mean should ignore local variation in the spread of a spatial distribution. The influence of unusually high or low values in the locality may move the position of the mean if there were a few unusually expensive properties among others of relatively low cost. One possibility for a locally weighted variance might be

$$\sigma^2(z) = \frac{\sum_{i=1}^{n} w_i(z)[x_i - \tilde{x}(z)]^2}{\sum_{i=1}^{n} w_i(z)} \tag{10.3}$$

The deviances calculated in the numerator are relative to the local mean calculated by Equation (10.2). Examining variation in the local variance will give some indication of differences in the local spread of the distribution across the map. Again, a map of this statistic may be examined in conjunction with a map of local means. By extension, formulae for the local skewness, local covariance and local Pearson product moment correlation coefficient may also be derived.

Another extension of the GWR concept is to the range of multivariate statistical techniques other than regression which have frequently been employed by geographers. This includes principal components analysis (and, by extension, factor analysis) and discriminant analysis. The assumption made in their use is that the model coefficients are spatially invariant, and that any spatial variation can be safely confined to the error term. However, if spatial variation is important, treating it as an error is perhaps rather perverse.

Any of these methods for which continuous observation weighting is possible may be subjected to geographical weighting in the manner of GWR. However, the analyst doing so is presented with a daunting array of results. With geographically weighted principal components, there will be a set of component loadings and component scores for every location in the study area for which components are extracted. This does present a problem with interpretation, since the component which accounts for the greatest amount of local variance in one part of the study may not be directly comparable with the component which accounts for the greatest amount of local variance elsewhere. Certainly one may map the coefficients, but interpretation of the components becomes something of a daunting prospect. How does the analyst deal with spatially varying eigenvalues for example? Principal components analysis represents a transformation from one space in which the axes may not be orthogonal to one in which the axes are orthogonal. What geometrical interpretation may be attached to geographically weighted principal components? Can we use the component scores to help understand geographical processes? How do we decide the 'best' value for the bandwidth?

With geographically weighted discriminant analysis, the picture becomes more complex. Typically the output from a discriminant analysis comprises the coefficients for the discriminant functions, predictions of the group membership based on those functions, and a confusion matrix indicating the difference between prior and posterior group membership. If we apply geographical weighting techniques here, then we are presented with all this for every location at which the functions are extracted. Can the coefficients be meaningfully mapped? Can the functions be compared? It may be helpful to be able to explore where a particular variable is important in influencing group membership and where it is not. How does the analyst interpret the confusion matrices? Perhaps a frequency analysis of how many times group membership of an observation is correctly predicted will reveal those for which prediction is less influenced by spatial non-stationarity, and those for which the opposite is the case. What does this tell us about the characteristics of individual obervations?

10.2.3   Alternative inferential frameworks (Bayes, MCMC)

Generations of geography students have been taught to carry out classical statistical inference tests where a hypothesis is stated, a significance level is expressed, a test statistic calculated and compared with some value which relates to the desired significance level, and a decision is made to accept or reject the hypothesis. It is not always clear where the hypothesis comes from, nor how the significance level is arrived at. We end up testing 'pinhead' hypotheses: is it helpful to know that the regression parameter is not significantly different from zero at the 5% significance level? The problem is that we do not consider the question 'How different from zero is the regression parameter?' In the large spatial data sets increasingly available, even minute differences from zero (say 0.001) would be 'significantly' different from zero.

A notable trend in statistics over recent years has been the rise in the use of Bayesian inference (see Chapter 8). There are a number of reasons why this may have occurred. Among statisticians there has been a growing current of dissatisfaction with the classical 'significance testing' approach to statistical inference. Also, the development of Monte Carlo Markov chain (MCMC) techniques has made the Bayesian approach viable in many situations where it would otherwise not be (see for example Gelman et al., 1995). Essentially, MCMC techniques enable the properties of multivariate probability distributions to be investigated by simulation, in situations when analytical approaches are not practical. The number of simulations needed is sometimes very large (in excess of a million in some situations), but at the time of writing this does not present a challenge to even fairly standard computing equipment. Analysts are far less constrained by the mathematical properties of the statistical models they may apply within the MCMC framework, since it is no longer necessary to specify models that are 'mathematically convenient'. Thus models that are in some sense more realistic may be specified and results of greater utility may be obtained.

This is of particular importance to quantitative geographers, since there is growing interest in Bayesian MCMC approaches to spatial statistical problems (Besag and Green, 1993). Spatial statistics are notorious for generating analytically intractable problems and so MCMC techniques could prove to be a very useful innovation for this subject matter.

10.2.4 Geometry

It is perhaps the historical prevalence of paper maps which has led us to assume that calculations relating to the surface of the earth should take place using a Cartesian coordinate system on a plane. While this is acceptable for small-scale processes, increasingly we are interested in global processes and global models of such processes where the spherical nature of the earth needs to be considered. Does thinking about the earth as a sphere provide any opportunities? We suggested in Chapter 2 that calculations using spherical coordinates are rather cumbersome. However, the situation is less of a problem than might be imagined (Brannan et al., 1999). A point on the surface of the earth can be thought of as being in three-dimensional Euclidean space with coordinates (x, y, z). The centre of the earth is taken to the point (0,0,0). If we assume the radius of the earth to be unity, then the positive x axis meets the surface where the Greenwich meridian crosses the equator at (1,0,0). The positive y axis meets the surface where the 90°E meridian crosses the equator at (0,1,0), and the z axis meets the surface at (0,0,1). The point (0,0,1) is the North Pole, and (0,0,−1) is the South Pole. If a plane cuts the surface of the earth and passes through the origin, the circle formed by the intersection of the plane and the earth's surface is known as a great circle. Great circles that pass through the North Pole are known as meridians; that passing through (1,0,0) is known as the Greenwich meridian. The equator is a great circle, but other lines of latitude are not.

If we know the latitude θ and longitude λ of a point, the spherical coordinates are given by $(\cos \lambda \sin(\pi/2 - \theta), \sin \lambda \sin(\pi/2 - \theta), \cos(\pi/2 - \theta))$. If the coordinates of two locations are known, say (a_1, a_2, a_3) and (b_1, b_2, b_3), the distance between the two points is given by

$$d_{12} = R \times \text{arc} \cos(a_1 b_1 + a_2 b_2 + a_3 b_3) \qquad (10.4)$$

where R is the radius of the earth. Pythagoras' theorem has a spherical analogue. If ABC are the vertices of a triangle on the surface of the earth, with the right angle at C, then the relationship between the lengths is

$$AB = \text{arc} \cos(\cos BC + \cos CA) \qquad (10.5)$$

where AB is the length of the side between the vertices A and B calculated from Equation (10.5). For any triangle on the sphere, where the angle at the vertex C is γ, the cosine rule can be used:

$$AB = \text{arc} \cos(\cos BC \cos CA + \sin BC \sin CA \cos \gamma) \qquad (10.6)$$

We have become used to thinking of spatial data processing as largely two dimensional and Euclidean. Yet, for continental applications, or global applications, this presents some difficulties. We can invert our normal practice, store our data with spherical coordinates, and carry out the geometrical operations using spherical methods. We can then project the results onto a plane when we require some sort of display in two dimensions; in this way, we may then start to understand more about the spatial processes operating at a global scale on the earth's surface. An interesting example is provided by Schettino (1999) who considers methods for polygon intersection in spherical topology. He uses these to model the historical movement of tectonic plates on the earth's surface.

Other applications where non-Euclidean geometries would seem to be appropriate concern individuals' perceptions of distances between places. It is highly unlikely that perceived or cognized distances follow basic Euclidean rules of geometry yet such distances are likely to play a major role in affecting people's spatial behaviour.

10.2.5   Basic spatial concepts: proximity and accessibility

Several techniques in spatial analysis, such as the classical measures of spatial autocorrelation, rely upon a matrix of proximity measures between zones where each proximity measure represents the degree of influence of one zone upon another. We suspect that with some spatial processes there is a 'contagion' effect and hence the value of an attribute in one zone can have an effect on the values of that attribute in neighbouring zones. The problem arises in that there is no one acceptable definition of what constitutes a 'neighbouring' zone. Proximity is sometimes defined in a discrete manner so that the degree of connection between two zones is one if the two zones share a common boundary and zero otherwise. Sometimes, it is measured in a continuous manner, perhaps as a function of the length of the common boundary or the inverse distance between the centroids of the zones or some other, more complex, function (Bavaud, 1998). However, various problems can still arise. Suppose the common boundary between two zones is a river, and there is little interaction between the two zones (due to lack of bridging points, perhaps). How does this change our view of contiguity? If there are no links, is there no contiguity? If there is one link in a 100 km border, is there different contiguity from 50 links in a border of the same length? If two zones meet at a point, can they be said to be contiguous?

Clearly all these ways of measuring spatial proximity have some validity and there appears to be no one commonly accepted measure. The problem arises in the spatial analytical techniques that incorporate such a measure – a change in the definition of spatial proximity can have a dramatic effect on the resulting analysis. It therefore makes the calculation of statistics such as spatial autocorrelation highly subjective. The paradox is that we berate those who do not use techniques such as spatial regression models, which involve a measure of spatial autocorrelation, because of the possibility of producing incorrect standard errors of the parameter estimates, yet the solution we offer depends on a highly subjective definition of spatial proximity. As noted in Chapter 7, the subjectivity of the definition of spatial proximity in such techniques must add to the uncertainty we have about the true nature of the parameter estimates but this uncertainty is not recognized in any statistical way.

A similar problem arises in the definition of another basic spatial concept – that of accessibility. We are generally aware of what we mean when we say 'this house is accessible to the shops', but how do we measure such a concept consistently? For example, one commonly used measure of accessibility is the potential measure attributed to Hansen (1959):

$$A_i = \Sigma_j S_j^\alpha d_{ij}^\beta \tag{10.7}$$

where A_i is the accessibility of place i for some activity such as shopping, S_j is a measure of the number of opportunities, such as the square footage of shopping at j, and d_{ij} represents the distance between i and j. The parameters α and β reflect the importance of size and distance, respectively, on the determination of accessibility. The problem is that often there is no way to determine what these parameters are and variations in them can have significant effects on the accessibility surfaces produced. For example, the more negative is β, the more 'spiky' the accessibility surface becomes. Quite often, in the use of this measure of accessibility, α is assumed to be 1 and β is assumed to be -1, but there is usually no justification for using these particular values other than they produce 'reasonable looking' surfaces which are neither too smooth nor too spiky. Despite the attention of researchers such as Weibull (1976; 1980), Pooler (1987) and Miller (1999) to this problem, it probably warrants further investigation.

10.2.6 Merging space and time

A traditional adage has it that 'geographers study space and economists study time'. However, it is increasingly recognized that for some processes, a study of space *and* time is important: examples include the modelling of epidemics (Seimiatycki et al., 1980) and the analysis of patterns of household burglaries (Brunsdon, 1989). Typical analytical approaches test the hypothesis of 'space–time' interaction; that is, they examine whether the locations of events and the points in time at which they occur are independent or not. For example, space–time interaction in household burglaries might imply that certain streets are targeted at certain times. Note that space–time interaction implies more than just spatial clustering (some places experience more burglaries than others regardless of time) or just temporal clustering (the overall risk of burglary varies by season). In epidemiology, testing for space–time interaction is particularly important because the presence of such interaction implies that some degree of contagion is involved in the occurrence of a disease.

The earliest, and perhaps simplest, test for space–time interaction was suggested by Knox (1964) and elaborated on by Mantel (1967). However, both tests suffer the problem that an arbitrary definition of 'closeness' in space and time has to be made. In the Knox test, this is in terms of binary indicators of closeness in space and time; in the Mantel test, it is in terms of coefficients in continuous measures of closeness. More recently, a K function approach (see Chapter 6) to the detection of space–time interaction has been proposed by Diggle et al. (1993) and Bailey and Gatrell (1995).

As with purely spatial patterns, the opportunity to explore and model patterns in both space and time has greatly increased with the advent of powerful desktop computers and this area holds much potential for further developments. For example, MCMC-based methods can be used to consider stochastic space–time processes (Knorrheld and Besag, 1998) or to carry out bootstrap estimations of space–time K functions. Another promising area for space–time data is that of visualization, and again, with the advent of ever more powerful computing environments with strong graphics capabilities, techniques such as movie making allow the direct visualization of space–time processes (Dorling and Openshaw, 1992).

10.3 Training people to think spatially

10.3.1 Teaching quantitative geography

Many university degree courses in geography provide a 'quantitative methods' module introducing descriptive statistics, 'simple' statistical analysis, and an introduction to classical (Neyman–Pearson) statistical tests. Many students find the nature of conducting a statistical test confusing and problematic. Such courses usually begin with elementary descriptive measures, proceed through elementary probability theory and inference, and through a range of classical methods, including correlation and simple linear regression, perhaps with a dash of non-parametric statistics thrown in for good measure. There may be a follow-on course that includes a selection from further regression topics, principal components and factor analysis, discriminant analysis, analysis of variance, and cluster analysis. Ecologists may be keen to include methods based on correspondence analysis. In all of this, we run the risk of losing sight of the many special aspects of spatial data. Courses dealing explicitly with spatial modelling and spatial statistics are relatively rare. Is there a need for us to rethink the way in which we approach the teaching of quantitative techniques in geography? Should we introduce the role of space very early into quantitative courses and concentrate more heavily on the types of spatial techniques described in this book?

A similar set of questions relate to how we persuade colleagues, whose academic familiarity with quantitative techniques is either negligible or harks back to a distant era and is now out of date, that modern quantitative geography has something very useful to offer both themselves and their students. As Wilson (1984, p. 212) notes:

> much of the criticism of 'quantitative geography' ... is directed at the work of the 1950s and 1960s (a lot of this style of work is still taught and still appears in the journals); many of the critics do not have a picture of the steady, indeed often dramatic, progress which has been made since the 1950s.

There is no doubt that many of the new methods require an understanding of the language of mathematics, but not necessarily at advanced level. Experience suggests that undergraduates can understand the nature and utility of GWR, for example, and that they appreciate the essential role that space plays in both the determination of the statistics and their interpretation.

10.3.2   Software

Most academic computer centres provide a standard range of statistical software for data analysis. Few of these have any appreciable facilities for the analysis of spatial data at the time of writing. In order to popularize the use of purely spatial statistics such as those described here, it would be useful to have a module of SAS or SPSS devoted to spatial analysis. Performing routines such as GWR would then become commonplace and we could forget about teaching inappropriate techniques simply because they are the only ones readily available in a statistics package. With the increasing availability of spatial data and the increasing recognition that spatial data have different properties from aspatial data, there will be an increasing demand for statistical packages which can handle spatial data analysis.

The role of GIS could be crucial here. It is not always clear that integrating prepackaged spatial analysis methods inside GIS software constitutes an ideal way forward (Fotheringham, 1999b). A large proportion of current spatial analysis routines are not fully implemented with GIS, reflecting the fact that the purpose of most commercial GIS is focused on data handling and management, rather than on data analysis. Dynamic linkages between software may well be a more promising route. However, there is no denying that the power and proliferation of GIS has led to a dramatic increase in the recognition that spatial data demand specialized techniques. It may well be that GIS do become the equivalent of a word processor for spatial data and that within a decade, advanced spatial analysis routines will be standard features of such systems.

10.4   Summary

The message of this book is hopefully a fairly clear one: spatial data have special properties that create a need for specialized techniques for both spatial data analysis and spatial modelling. The increasing recognition of this fact has led to a reduced reliance on techniques developed in aspatial disciplines and which have been transplanted to subjects such as geography and applied to spatial data. It has promoted an increased awareness of the importance of developing specifically *spatial* techniques that can handle the unique properties of the subject matter. It has also brought forth a renewed interest in the role that space plays in various processes and, in particular, the role that quantitative analysis can play in understanding these processes.

In a sense, quantitative geography is undergoing an internal revolution as we increasingly recognize the importance of space in providing a unique milieu for analysis and increasingly have the confidence to develop explicitly *spatial* theory. As Morrill (1984, p. 68) notes, writing 17 years ago:

> I assert that for a few people (going in many and varied directions), we have made considerable progress, both theoretically and practically, for so ridiculously short a time (20 years). We have barely started the process of construction and testing of geographic theory.

The thesis of this book is that we are now at the beginning of an era of renewed interest in quantitative spatial analysis with the development of explicitly spatial techniques. It is an era when advances are being made, not so much by borrowing from other disciplines, but by developing spatial theory based on knowledge of spatial processes. It is an era when the constraints are not so much to do with knowledge of mathematics or statistics but to do with one's imagination. The challenge is to think about space and spatial processes in new and exciting ways. It is an interesting time to be a quantitative geographer.

Bibliography

Aitkin, M. (1996) 'A General Maximum Likelihood Analysis of Overdispersion in Generalized Linear Models', *Statistics and Computing*, 6: 251–62.

Alker, H.S. (1969) 'A Typology of Ecological Fallacies', in M. Dogan and S. Rokkan (eds), *Quantitative Ecological Analysis*. Boston: MIT Press. pp. 64–86.

Alonso, W. (1978) 'A Theory of Movement', in N.M. Hansen (ed.), *Human Settlement Systems*. Cambridge: Ballinger. pp. 197–211.

Ankerst, M., Keim, D. and Kreigel, H.-P. (1996) 'Circle Segments: A Technique for Visually Exploring Large Multidimensional Data Sets', *I.E.E.E. Visualization '96 Proceedings*, Hot Topics Session, San Fancisco, CA.

Anselin, L. (1988) *Spatial Econometrics, Methods and Models*. Dordrecht: Kluwer Academic.

Anselin, L. (1992) *SpaceStat Tutorial: A Workbook for using SpaceStat in the Analysis of Spatial Data*, Technical Software Series S-92-1, National Center for Geographic Information and Analysis, Santa Barbara, CA.

Anselin, L. (1995) 'Local Indicators of Spatial Association – LISA', *Geographical Analysis*, 27: 93–115.

Anselin, L. (1996) 'The Moran Scatterplot as an ESDA Tool to Assess Local Instability in Spatial Association', in M.M. Fischer, H. Scholten and D. Unwin (eds), *Spatial Analytical Perspectives on GIS*. London: Taylor and Francis.

Anselin, L. (1998) 'Exploratory Spatial Data Analysis in a Geocomputational Environment', in P.A. Longley, S.M. Brooks, R. McDonnell and B. Macmillan (eds), *Geocomputation: A Primer*. Chichester: Wiley. pp. 77–94.

Anselin, L. and Bao, S. (1997) 'Exploratory Spatial Data Analysis Linking SpaceStat and ArcView', in M.M. Fischer and A. Getis (eds), *Recent Developments in Spatial Analysis*. Berlin: Springer-Verlag. pp. 35–59.

Anselin, L. and Rey, S.J. (1997) 'Introduction to the Special Issue on Spatial Econometrics', *International Regional Science Review*, 20: 1–8.

Asimov, D. (1985) 'The Grand Tour: A Tool for Viewing Multidimensional Data', *SIAM Journal on Scientific and Statistical Computing*, 6: 128–43.

Aten, B. (1997) 'Does Space Matter? International Comparisons of the Prices of Tradeables and Non-Tradeables', *International Regional Science Review*, 20: 35–52.

Atkins, D.J. and Fotheringham, A.S. (1999) 'Gender Variations in Migration Destination Choice', in P. Boyle and K. Halfacree (eds), *Migration and Gender in the Developed World*. London: Routledge. pp. 54–72.

Bailey, T.C. and Gatrell, A.C. (1995) *Interactive Spatial Data Analysis*. Harlow: Longman.

Bao, S. and Henry, M. (1996) 'Heterogeneity Issues in Local Measurements of Spatial Association', *Geographical Systems*, 3: 1–13.

Barndorff-Nielsen, O., Jensen, J. and Kendall, W.S. (1997) *Networks and Chaos: Statistical and Probabilistic Aspects*. London: Chapman and Hall.

Bartholomew (no date) Bartholomew Digital Map Data, *http://www.geo.ed.ac.uk/~barts_twr/framdig.html*, accessed on 21 March 1999.

Batsell, R.R. (1981) 'A Multiattribute Extension of the Luce Model which Simultaneously Scales Utility and Substitutability', Working Paper, J.H. Jones Graduate School of Administration, Rice University, Texas.

Baxter, R.S. (1976) *Computer and Statistical Techniques for Planners*. London: Methuen.

Bavaud, F. (1998) 'Models for Spatial Weights: A Systematic Look', *Geographical Analysis*, 30: 153–71.

Bayes, T. (1763) 'An Essay Towards Solving a Problem in the Doctrine of Chances', *Philosophical Transactions of the Royal Society*, 330–418.

Berry, D. (1997) 'Teaching Elementary Bayesian Statistics with Real Applications in Science', *The American Statistician*, 51: 241–6.

Berry, J. (1998) 'Unlock the Statistical Keystone', *Geoworld*, 11: 28–9.

Berry, W.D. and Feldman, S. (1985) *Multiple Regression in Practice*, Quantitative Applications in the Social Sciences, 50. London: Sage.

Besag, J. and Green, P. (1993) 'Spatial Statistics and Bayesian Computation', *Journal of the Royal Statistical Society*, B, 55: 25–37.

Besag, J. and Newell, J. (1991) 'The Detection of Clusters in Rare Diseases', *Journal of the Royal Statistical Society*, A, 154: 143–55.

Bettman, J.R. (1979) *An Information Processing Theory of Consumer Choice*. Reading, MA: Addison-Wesley.

Bickel, P. and Freedman, D. (1981) 'Some Asymptotics on the Bootstrap', *Annals of Statistics*, 9: 1196–217.

Billinge, M., Gregory, D. and Martin, R. (eds) (1984) *Recollections of a Revolution*. New York: St. Martin's Press.

Bithell, J. and Stone, R. (1989) 'On Statistical Methods for Analysing the Geographical Distribution of Cancer Cases near Nuclear Installations', *Journal of Epidemiology and Community Health*, 43: 79–85.

Blakemore, M.J. (1984) 'Generalisation and Error in Spatial Data Bases', *Cartographica*, 21: 131–9.

Blalock, H. (1964) *Causal Inferences on Nonexperimental Research*. Chapel Hill, NC: University of North Carolina Press.

Bonham-Carter, G. (1994) *Geographic Information Systems for Geoscientists*. Oxford: Pergamon.

Boots, B. and Getis, A. (1988) *Point Pattern Analysis*. London: Sage.

Bowman, A. and Azzelini, A. (1997) *Applied Smoothing Techniques for Data Analysis*. Oxford: Oxford University Press.

Bowman, A.W. (1984) 'An Alternative Method of Cross-Validation for the Smoothing of Density Estimates', *Biometrika*, 71: 353–60.

Bowyer, A. and Woodwark, J. (1983) *A Programmer's Geometry*. London: Butterworth.

Boyle, P.J., Flowerdew, R. and Shen, J. (1998) 'Analysing Local-Area Migration Using Data from the 1991 Census: The Importance of Housing Growth and Tenure', *Regional Studies*, 32: 113–32.

Bracken, I. and Martin, D. (1989) 'The Generation of Spatial Population Distributions from Census Centroid Data', *Environment and Planning A*, 21: 537–43.

Bradley, W.J. and Schaefer, K.C. (1998) *The Uses and Misuses of Data and Models: The Mathematization of the Human Sciences*. London: Sage.

Brannan, D.A., Esplan, M. and Gray, J.J. (1999) *Geometry*. Cambridge: Cambridge University Press.

Bras, R. and Rodriguez-Iturbe, I. (1985) *Random Functions and Hydrology*. Reading, MA: Addison-Wesley.

Bresenham, J.E. (1965) 'Algorithm for the Computer Control of a Digital Plotter', *IBM Systems Journal*, 4: 25–30.

Broomhead, D. and Lowe, D. (1988) 'Multivariable Functional Interpolation and Adaptive Networks', *Complex Systems*, 2: 321–55.

Brown, L.A. and Goetz, A.R. (1987) 'Development Related Contextual Effects and Individual Attributes in Third World Migration Processes: A Venezuelan Example', *Demography*, 24: 497–516.

Brown, L.A. and Jones III, J.P. (1985) 'Spatial Variation in Migration Processes and Development: A Costa Rican Example of Conventional Modeling Augmented by the Expansion Method', *Demography*, 22: 327–52.

Brown, L.A. and Kodras, J.E. (1987) 'Migration, Human Resources Transfer and Development Contexts: A Logit Analysis of Venezuelan Data', *Geographical Analysis*, 19: 243–63.

Brunsdon, C. (1989) 'Spatial Analysis Techniques Applied to Local Crime Patterns', PhD Thesis, Department of Geography, University of Newcastle.

Brunsdon, C. (1995a) 'Analysis of Univariate Census Data', in S. Openshaw (ed.), *Census Users Handbook*. Cambridge: GeoInformation International. pp. 213–38.

Brunsdon, C. (1995b) 'Estimating Probability Surfaces for Geographical Points Data: An Adaptive Kernel Algorithm', *Computers and Geosciences*, 21: 877–94.

Brunsdon, C. and Charlton, M.E. (1996) 'Developing an Exploratory Spatial Analysis System in XlispStat', in D. Parker (ed.), *Innovations in GIS 3*. London: Taylor and Francis. pp. 135–45.

Brunsdon, C., Fotheringham, A.S. and Charlton, M.E. (1996) 'Geographically Weighted Regression: A Method for Exploring Spatial Nonstationarity', *Geographical Analysis*, 28: 281–98.

Brunsdon, C., Fotheringham, A.S. and Charlton, M.E. (1998a) 'Spatial Non-Stationarity and Autoregressive Models', *Environment and Planning A*, 30: 957–73.

Brunsdon, C., Fotheringham, A.S. and Charlton, M.E. (1998b) 'Geographically Weighted Regression – Modelling Spatial Non-stationarity', *The Statistician*, 47: 431–43.

Brunsdon, C., Fotheringham, A.S. and Charlton, M.E. (1999a) 'Some Notes on Parametric Significance Tests for Geographically Weighted Regression', *Journal of Regional Science*, 39: 497–524.

Brunsdon, C., Aitkin, M., Fotheringham, A.S. and Charlton, M.E. (1999b) 'A Comparison of Random Coefficient Modelling and Geographically Weighted Regression for Spatially Non-Stationary Regression Problems', *Geographical and Environmental Modelling*, 3: 47–62.

Bugayevskiy, L.M. and Synder, J.P. (1995) *Map Projections: A Reference Manual*. London: Taylor and Francis.

Burrough, P.A. (1986) *Principles of Geographical Information Systems for Land Resources Assessment*. Oxford: Oxford University Press.

Burrough, P.A. (1996) 'Natural Objects with Indeterminate Boundaries', in P.A. Burrough and A.U. Frank (eds), *Geographic Objects with Indeterminate Boundaries*. London: Taylor and Francis. pp. 3–28.

Burrough, P.A. and Frank, A.U. (1996) *Geographic Objects with Indeterminate Boundaries*. London: Taylor and Francis.

Burrough, P.A. and McDonnell, R.A. (1998) *Principles of Geographical Information Systems*. Oxford: Oxford University Press.

Burton, I. (1963) 'The Quantitative Revolution and Theoretical Geography', *The Canadian Geographer*, 7: 151–62.

Carey, H.C. (1858) *Principles of Social Science*, Vol. 1. Philadelphia: Lippincott.

Carpenter, G. and Grossberg, S. (1988) 'The ART of Adaptive Pattern Recognition by a Self-organizing Network', *Computer*, 21: 77–88.

Carver, S.J. and Brunsdon, C. (1994) 'Vector to Raster Conversion and Feature Complexity:

An Empirical Study using Simulated Data', *International Journal of Geographical Information Systems*, 8: 261–70.

Casetti, E. (1972) 'Generating Models by the Expansion Method: Applications to Geographic Research', *Geographical Analysis*, 4: 81–91.

Casetti, E. (1982) 'Drift Analysis of Regression Parameters: An Application to the Investigation of Fertility Development Relations', *Modeling and Simulation*, 13: 961–6.

Casetti, E. (1997) 'The Expansion Method, Mathematical Modeling, and Spatial Econometrics', *International Regional Science Review*, 20: 9–32.

Casetti, E. and Jones III, J.P. (1983) 'Regional Shifts in the Manufacturing Productivity Response to Output Growth: Sunbelt versus Snowbelt', *Urban Geography*, 4: 286–301.

Charlton, M.E., Fotheringham, A.S. and Brunsdon, C. (1996) 'The Geography of Relationships: An Investigation of Spatial Non-Stationarity', in J.-P. Bocquet-Appel, D. Courgeau and D. Pumain (eds), *Analyse Spatiale de Données Biodémographiques*. Montrouge: John Libbey Eurotext. pp. 23–47.

Chorley, R.J., Stoddart, D.R., Haggett, P. and Slaymaker, O. (1966) 'Regional and Local Components in the Areal Distribution of Surface Sand Faces in the Breckland, E. England', *Journal of Sedimentary Petrology*, 36: 209–20.

Chrisman, N. (1997) *Exploring Geographic Information Systems*. Chichester: Wiley.

Christaller, W. (1933) *Die Zentralen Orte in Süd-Deutschland*. Jena: Gustav Fischer Verlag.

Clarke, D.E. and Herrin, W.E. (1997) 'Public School Quality and Home Sale Prices: Evidence from a California Housing Market', Paper presented at the 44th Annual Meeting of the America Regional Science Association, Buffalo, NY.

Clarke, K.C. (1997) *Getting Started with Geographic Information Systems*. Englewood Cliffs, NJ: Prentice Hall.

Cleveland, W.S. (1979) 'Robust Locally Weighted Regression and Smoothing Scatterplots', *Journal of the American Statistical Association*, 74: 829–36.

Cleveland, W.S. (1993) *Visualizing Data*. Summit, NJ: Hobart Press.

Cleveland, W.S. and Devlin, S.J. (1988) 'Locally Weighted Regression: An Approach to Regression Analysis by Local Fitting', *Journal of the American Statistical Association*, 83: 596–610.

Cliff, A.D. and Ord, J.K. (1973) *Spatial Autocorrelation*. London: Pion.

Cliff, A.D. and Ord, J.K. (1981) *Spatial Processes: Models and Applications*. London: Pion.

Codd, E.F. (1970) 'A Relational Model of Data for Large Shared Data Banks', *Communications of the Association for Computing Machinery*, 13: 377–87.

Cook, D. and Buja, A. (1997) 'Manual Controls for High Dimensional Data Projections', *Journal of Computational and Graphical Statistics*, 6: 464–80.

Cook, D. and Pocock, S. (1983) 'Multiple Regression in Geographical Mortality Studies, with Allowance for Spatially Correlated Errors', *Biometrics*, 39: 361–71.

Cook, D., Buja, A. and Swayne, D.F. (1993) 'Projection Pursuit Indexes Based on Orthonormal Function Expansions', *Journal of Computational and Graphical Statistics*, 2: 225–50.

Cook, D., Buja, A. Cabrera, J. and Hurley, C. (1995) 'Grand Tour and Projection Pursuit', *Journal of Computational and Graphical Statistics*, 4: 155–72.

Cook, D., Cruz-Neira, C., Kohlmeyer, B.D., Lechner, U., Lewin, N., Nelson, L., Olsen, A., Pierson, L. and Symanzyk, J. (1998) 'Exploring Environmental Data in a Highly Immersive Virtual Reality Environment', *Environmental Monitoring and Assessment*, 51: 441–50.

Cook, D., Majure, J.J., Symanzik, J. and Cressie, N. (1996) 'Dynamic Graphics in a GIS: Exploring and Analyzing Multivariate Spatial Data using Linked Software', *Computational Statistics*, 11: 467–80.

Cook, D., Symanzik, J., Majure, J.J. and Cressie, N. (1997) 'Dynamic Graphics in a GIS: More Examples using Linked Software', *Computers and Geosciences*, 23: 371–85.

Cooley, W.W. and Lohnes, P.R. (1962) *Multivariate Procedures for the Behavioral Sciences.* New York: Wiley.

Cressie, N. (1984) 'Towards Resistant Geostatistics', in G. Verly, M. David, A.G. Journel and A. Marachel (eds), *Geostatistics for Natural Resources Characterization.* Dordrecht: Reidel, Part 1: 21–44.

Cressie, N. (1993) *Statistics for Spatial Data.* New York: Wiley.

Curtis, A. and Fotheringham, A.S. (1995) 'Large Scale Information Surfaces: An Analysis of City Name Recalls in the United States', *Geoforum*, 26: 65–78.

Dacey, M.F. (1960) 'The Spacing of River Towns', *Annals of the Association of American Geographers*, 50: 59–61.

Dale, A. and Marsh, C. (1993) *The 1991 Census User's Guide.* London: HMSO.

DeGroot, M. (1988) *Probability and Statistics*, 2nd edn. Redwood City, CA: Addison-Wesley.

DeMers, M. (1997) *Fundamentals of Geographic Information Systems.* Chichester: Wiley.

Dempster, A.P., Laird, N.M. and Rubin, D.B. (1977) 'Maximum Likelihood from Incomplete Data via the EM Algorithm', *Journal of the Royal Statistical Society*, Series B, 39: 1–38.

Diaconis, P. and Efron, B. (1983) 'Computer Intensive Methods in Statistics', *Scientific American*, 248: 116–30.

Diaconis, P. and Freedman, D. (1984) 'Asymptotics of Graphical Projection Pursuit', *The Annals of Statistics*, 12: 793–815.

Diggle, P. (1983) *The Statistical Analysis of Point Patterns.* London: Academic Press.

Diggle, P.J., Chetwynd, A.G., Hagqvist, R. and Morris, S. (1993) 'Second-Order Analysis of Space-Time Clustering', Technical Report, Department of Mathematics, University of Lancaster.

Diggle, P., Gatrell, A. and Lovett, A. (1990) 'Modelling the Prevalence of Cancer of the Larynx in Part of Lancashire: A New Methodology for Spatial Epidemiology', in R.W. Thomas (ed.), *Spatial Epidemiology.* London: Pion. pp. 35–47.

Diggle, P., Tawn, J. and Moyeed, A. (1998) 'Model-Based Geostatistics', *Applied Statistics*, 47: 299–350.

Dijkstra, E.W. (1959) 'A Note on Two Problems in Connection with Graphs', *Numerische Mathematik*, 1: 269–71.

Ding, Y. and Fotheringham, A.S. (1992) 'The Integration of Spatial Analysis and GIS', *Computers, Environment and Urban Systems*, 16: 3–19.

Dobson, A. (1990) *An Introduction to Generalized Linear Models.* London: Chapman and Hall.

Dodd, S.C. (1950) 'The Interactance Hypothesis: A Model Fitting Physical Masses and Human Groups', *American Sociological Review*, 15: 245–57.

Doll, R. (1989) 'The Epidemiology of Childhood Leukaemia', *Journal of the Royal Statistical Society*, A, 152: 341–51.

Dorling, D.F.L. (1991) 'The Visualization of Spatial Social Structure'. PhD Thesis, University of Newcastle upon Tyne.

Dorling, D.F.L. (1995) *A New Social Atlas of Britain.* Chichester: Wiley.

Dorling, D.F.L. and Openshaw, S. (1992) 'Using Computer Animation to Visualise Space-Time Patterns', *Environment and Planning B*. 19: 639–50.

Downs, R.M. and Stea, D. (1977) *Maps in Minds: Reflections on Cognitive Mapping.* New York: Harper and Row.

Dowson, R.M. and Wragg, A. (1973) 'Maximum-entropy Distributions having prescribed First and Second Moments', *IEEE Transactions on Information Theory*, IT-19: 689–93.

Draper, N. and Smith, H. (1981) *Applied Regression Analysis.* New York: Wiley.

Duncan, C. (1997) 'Applying Mixed Multivariate Multilevel Models in Geographical Research', in G.P. Westert and R.N. Verhoeff (eds), *Places and People: Multilevel*

Modelling in Geographical Research, Nederlandse Geografische Studies 227, University of Utrecht. pp. 100–17.

Duncan, C., Jones, K. and Moon, G. (1996) 'Health-Related Behaviour in Context: A Multilevel Approach', *Social Science and Medicine*, 42: 817–30.

Efron, B. (1981) 'Nonparametric Standard Errors and Confidence Intervals' (with discussion), *Canadian Journal of Statistics*, 9: 139–72.

Efron, B. (1982) *The Jackknife, the Bootstrap and other Resampling Plans*. Philadelphia: Society for Industrial and Applied Mathematics.

Efron, B. and Gong, G. (1983) 'A Leisurely Look at the Bootstrap, the Jackknife and Cross-validation', *American Statistician*, 37: 36–48.

Efron, B. and Tibshirani, R. (1986) 'Bootstrap Methods for Standard Errors, Confidence Intervals, and other Measures of Statistical Accuracy', *Statistical Science*, l: 54–77.

Ehrenberg, A. (1982) *A Primer in Data Reduction*. Chichester: Wiley.

Eldridge, J.D. and Jones III, J.P. (1991) 'Warped Space: A Geography of Distance Decay', *Professional Geographer*, 43: 500–11.

Fik, T.J. and Mulligan, G.F. (1990) 'Spatial Flows and Competing Central Places: Towards a General Theory of Hierarchical Interaction', *Environment and Planning A*, 22: 527–49.

Fik, T.J., Amey, R.G. and Mulligan, G.F. (1992) 'Labor Migration Amongst Hierarchically Competing and Intervening Origins and Destinations', *Environment and Planning A*, 24: 1271–90.

Fischer, M.M. and Getis, A. (eds) (1997) *Recent Developments in Spatial Analysis*. Berlin: Springer-Verlag.

Fisher, R.A. and Tippett, L.H.C. (1928) 'Limiting Forms of the Frequency Distribution of the Largest or Smallest Number of a Sample'. *Proceedings of the Cambridge Philosophical Society*, 24: 180–90.

Flowerdew, R. (1988) 'Statistical Methods for Areal Interpolation: Predicting Count Data from a Binary Variable', Research Report 16, Northern Regional Research Laboratory, Universities of Lancaster and Newcastle.

Flowerdew, R. and Green, M. (1991) 'Data Integration: Statistical Methods for Transferring Data between Zonal Systems', in I. Masser and M. Blakemore (eds), *Handling Geographic Information*. Harlow: Longman. pp. 38–54.

Foster, S.A. and Gorr, W.L. (1986) 'An Adaptive Filter for Estimating Spatially Varying Parameters: Application to Modeling Police Hours Spent in Response to Calls for Service', *Management Science*, 32: 878–89.

Fotheringham, A.S. (1981) 'Spatial Structure and Distance-Decay Parameters', *Annals of the Association of American Geographers*, 71: 425–36.

Fotheringham, A.S. (1983a) 'Some Theoretical Aspects of Destination Choice and Their Relevance to Production-Constrained Gravity Models', *Environment and Planning A*, 15: 1121–32.

Fotheringham, A.S. (1983b) 'A New Set of Spatial Interaction Models: The Theory of Competing Destinations', *Environment and Planning A*, 15: 15–36.

Fotheringham, A.S. (1984) 'Spatial Flows and Spatial Patterns', *Environment and Planning A*, 16: 529–43.

Fotheringham, A.S. (1986) 'Modelling Hierarchical Destination Choice', *Environment and Planning A*, 18: 401–18.

Fotheringham, A.S. (1988) 'Consumer Store Choice and Choice Set Definition', *Marketing Science*, 7: 299–310.

Fotheringham, A.S. (1989) 'Consumer Store Choice and Retail Competition', in S.K. Reddy and L. Pellegrini (eds), *Retail Marketing Channels: Economic and Marketing Perspectives on Producer-Distributor Relationships*. New York: Routledge. pp. 234–57.

Fotheringham, A.S. (1991a) 'Statistical Modeling of Spatial Choice: An Overview', in

A. Ghosh and C. Ingene (eds), *Spatial Analysis in Marketing: Theory, Methods, and Applications*. Greenwich, CT: JAI Press. pp. 95–118.

Fotheringham, A.S. (1991b) 'Migration and Spatial Structure: The Development of the Competing Destinations Model', in J. Stillwell and P. Congdon (eds), *Migration Models: Macro and Micro Approaches*. London: Bellhaven. pp. 57–72.

Fotheringham, A.S. (1992) 'Exploratory Spatial Data Analysis and GIS', *Environment and Planning A*, 24: 1675–8.

Fotheringham, A.S. (1994) 'On the Future of Spatial Analysis: The Role of GIS', *Environment and Planning A*, Anniversary Issue: 30–4.

Fotheringham, A.S. (1997a) 'Trends in Quantitative Methods I: Stressing the Local', *Progress in Human Geography*, 21: 88–96.

Fotheringham, A.S. (1997b) 'Geographic Information Systems: A New(ish) Technology for Statistical Analysis', in A. Unwin (ed.), *New Techniques and Technologies for Statistics II*. Amsterdam: IOS Press. pp 141–7.

Fotheringham, A.S. (1998) 'Trends in Quantitative Methods II: Stressing the Computational', *Progress in Human Geography*, 22: 283–92.

Fotheringham, A.S. (1999a) 'Geocomputational Analysis', in S. Openshaw, R.J. Abrahart and T.E. Harris (eds), *Geocomputation*. London: Taylor and Francis, at press.

Fotheringham, A.S. (1999b) 'GIS-Based Spatial Modelling: A Step Forwards or a Step Backwards?', in A.S. Fotheringham and M. Wegener (eds), *Spatial Models and GIS: A European Perspective*. London: Taylor and Francis. pp. 21–30.

Fotheringham, A.S. (1999c) 'Trends in Quantitative Methods III: Stressing the Visual', *Progress in Human Geography*, 23: 617–626.

Fotheringham, A.S. and Brunsdon, C. (1999) 'Local Forms of Spatial Analysis' *Geographical Analysis*, 31: 340–358.

Fotheringham, A.S. and Charlton, M.E. (1994) 'GIS and Exploratory Spatial Data Analysis: An Overview of Some Research Issues', *Geographical Systems*, 1: 315–27.

Fotheringham, A.S. and Curtis, A. (1992) 'Encoding Spatial Information: The Evidence for Hierarchical Processing', in A.U. Frank, I. Campari and U. Formentini (eds), *Theories and Methods of Spatio-Temporal Reasoning in Geographic Space*, Lecture Notes in Computer Science. Dortmund: Springer-Verlag. pp. 269–87.

Fotheringham, A.S. and Curtis, A. (1999) 'Regularities in Spatial Information Processing: Implications for Modelling Destination Choice', *Professional Geographer*, 51: 227–39.

Fotheringham, A.S. and Dignan, T. (1984) 'Further Contributions to a General Theory of Movement', *Annals of the Association of American Geographers*, 74: 620–33.

Fotheringham, A.S. and O'Kelly, M.E. (1989) *Spatial Interaction Models: Formulations and Applications*. London: Kluwer Academic.

Fotheringham, A.S. and Pitts, T.C. (1995) 'Directional Variation in Distance-Decay', *Environment and Planning A*, 27: 715–29.

Fotheringham, A.S. and Rogerson, P.A. (1993) 'GIS and Spatial Analytical Problems', *International Journal of Geographic Information Systems*, 7: 3–19.

Fotheringham, A.S. and Rogerson, P.A. (eds) (1994) *Spatial Analysis and GIS*. London: Taylor and Francis.

Fotheringham, A.S. and Trew, R. (1993) 'Chain Image and Store Choice Modeling: The Effects of Income and Race', *Environment and Planning A*, 25: 179–96.

Fotheringham, A.S. and Wong, D. (1991) 'The Modifiable Areal Unit Problem in Multivariate Statistical Analysis', *Environment and Planning A*, 23: 1025–44.

Fotheringham, A.S. and Zhan, F. (1996) 'A Comparison of Three Exploratory Methods for Cluster Detection in Spatial Point Patterns', *Geographical Analysis*, 28: 200–18.

Fotheringham, A.S., Brunsdon, C. and Charlton, M.E. (1999) 'Scale Issues and Geographically Weighted Regression', in N. Tate (ed.), *Scale Issues and GIS*. Chichester: Wiley, at press.

Fotheringham, A.S., Brunsdon, C. and Charlton, M.E. (1998) 'Geographically Weighted Regression: A Natural Evolution of the Expansion Method for Spatial Data Analysis', *Environment and Planning A*, 30: 1905–27.

Fotheringham, A.S., Charlton, M.E. and Brunsdon, C. (1996) 'The Geography of Parameter Space: An Investigation into Spatial Non-Stationarity', *International Journal of Geographical Information Systems*, 10: 605–27.

Fotheringham, A.S., Charlton, M.E. and Brunsdon, C. (1997a) 'Two Techniques for Exploring Non-stationarity in Geographical Data', *Geographical Systems*, 4: 59–82.

Fotheringham, A.S., Charlton, M.E. and Brunsdon, C. (1997b) 'Measuring Spatial Variations in Relationships with Geographically Weighted Regression', in M.M. Fischer and A. Getis (eds), *Recent Developments in Spatial Analysis: Spatial Statistics, Behavioral Modeling and Computational Intelligence*. Berlin: Springer-Verlag. pp. 60–82.

Fotheringham, A.S., Densham, P.J. and Curtis, A. (1995) 'The Zone Definition Problem in Location-allocation Modelling', *Geographical Analysis*, 27: 60–77.

Friedman, J. and Tukey, J. (1974) 'A Projection Pursuit Algorithm for Exploratory Data Analysis', *IEEE Transactions on Computers*, 23: 881–9.

Gardner, M.J. (1989) 'Review of Reported Increases of Childhood Cancer Rates in the Vicinity of Nuclear Installations in the UK', *Journal of the Royal Statistical Society*, A, 152: 307–25.

Gatrell, A.C. (1983) *Distance and Space: A Geographical Perspective*. Oxford: Clarendon Press.

Gatrell, A.C., Bailey, T.C., Diggle, P.J. and Rowlingson, B. (1996) 'Spatial Point Pattern Analysis and its Application in Geographical Epidemiology', *Transactions of the Institute of British Geographers*, 21: 256–74.

Gehlke, C.E. and Biehl, H. (1934) 'Certain Effects of Grouping upon the Size of the Correlation Coefficient in Census Tract Material', *Journal of the American Statistical Association,* Supplement 29: 169–70.

Gelman, A., Carlin, J., Stern, H.S. and Rubin, D.B. (1995) *Bayesian Data Analysis*. London: Chapman and Hall.

Georgescu-Roegen, N. (1971) *The Entropy Law and the Economic Process*. Cambridge, MA: Harvard University Press.

Getis, A. and Ord, J.K. (1992) 'The Analysis of Spatial Association by Use of Distance Statistics', *Geographical Analysis*, 24: 189–206.

Ghosh, A. and Rushton, G. (eds) (1987) *Spatial Analysis and Location-Allocation Models*. New York: Van Nostrand Rheinhold.

Gilbert, E.J. (1958) 'Pioneer Maps of Health and Disease in England', *Geographical Journal*, 124: 172–83.

Gober, P., Glasmeier, A.K., Goodman, J.M., Plane, D.A., Stafford, H.A. and Wood, J.S. (1995) 'Employment Trends in Geography', *The Professional Geographer*, 47: 336–46.

Goddard, J.B. and Kirby, A. (1976) *An Introduction to Factor Analysis*, Concepts and Techniques in Modern Geography, 7. Norwich: Geo Books.

Goldstein, H. (1987) *Multilevel Models in Educational and Social Research*. London: Oxford University Press.

Goldstein, H. (1994) 'Multilevel Cross-Classified Models', *Sociological Methods and Research*, 22: 364–75.

Goodchild, M.F. (1984) 'Geocoding and Geosampling', in G.L. Gaile and C.J. Willmott (eds), *Spatial Statistics and Models*. Dordrecht: Reidel. pp. 33–52.

Goodchild, M.F. (1986) *Spatial Autocorrelation*, Concepts and Techniques in Modern Geography, 47. Norwich: Geo Books.

Goovaerts, P. (1992) 'Factorial Kriging Analysis: A Useful Tool for Exploring the Structure of Multivariate Spatial Soil Information', *Journal of Soil Science*, 43: 597–619.

Goovaerts, P. (1999) 'Geostatistics in Soil Science: State-of-the-Art and Perspectives', *Geoderma*, 89: 1–45.

Gorr, W.L. and Olligschlaeger, A.M. (1994) 'Weighted Spatial Adaptive Filtering: Monte Carlo Studies and Application to Illicit Drug Market Modeling', *Geographical Analysis*, 26: 67–87.

Gould, M, (1996) 'What's so Special about Spatial?', *GIS Europe*, 10: 22.

Gould, P.R. (1970) 'Is Statistix Inferens the Geographical Name for a Wild Goose?', *Economic Geography*, 46: 439–48.

Gould, P.R. (1975) 'Acquiring Spatial Information', *Economic Geography*, 51: 87–99.

Gould, P.R. (1984) 'Statistics and Human Geography: Historical, Philosophical, and Algebraic Reflections', in G.L. Gaile and C.L. Wilmott (eds), *Spatial Statistics and Models*. Dordrecht: Reidel. pp. 17–32.

Gould, P.R. and White, R. (1974) *Mental Maps*. Boston: Allen & Unwin.

Graf, W. (1998) 'Why Physical Geographers Whine so Much', *The Association of American Geographers' Newsletter*, 33 (8): 2.

Graham, E. (1997) 'Philosophies Underlying Human Geography Research', in R. Flowerdew and D. Martin (eds), *Methods in Human Geography: A Guide for Students doing a Research Project*. Harlow: Longman. pp. 6–30.

Graybill, F.A. and Iyer, H.K. (1994) *Regression Analysis: Concepts and Applications*. Belmont, CA: Duxbury.

Greenwood, M.J. and Sweetland, D. (1972) 'The Determinants of Migration between Standard Metropolitan Statistical Areas', *Demography*, 9: 665–81.

Greig, D.M. (1980) *Optimisation*. London: Longman.

Greig-Smith, P. (1964) *Quantitative Plant Ecology*. London: Butterworth.

Griffith, D.A. (1987) *Spatial Autocorrelation: A Primer*. Washington, DC: Association of American Geographers.

Griffith, D.A. (1988) *Advanced Spatial Statistics*. Dordrecht: Kluwer Academic.

Grunsky, E.C. and Agterberg, F.P. (1992) 'Spatial Relationships of Multivariate Data', *Mathematical Geology*, 24: 731–58.

Gurney, K. (1995) *An Introduction to Neural Networks*. London: UCL Press.

Haines-Young, R., Green, D.R. and Cousins, S. (1993) *Landscape Ecology and GIS*. London: Taylor and Francis.

Haining, R.P. (1990) *Spatial Data Analysis in the Social and Environmental Sciences*. Cambridge: Cambridge University Press.

Haining, R.P., Wise, S. and Ma, J.S. (1998) 'Exploratory Spatial Data Analysis in a Geographic Information System Environment', *Journal of the Royal Statistical Society D: The Statistician*, 47: 457–69.

Hall, P. (1988) 'On Symmetric Bootstrap Confidence Intervals', *Journal of the Royal Statistical Society*, B, 50: 35–45.

Hansen, W.G. (1959) 'How Accessibility Shapes Land Use', *Journal of the American Institute of Planners*, 25: 73–6.

Harley, B.J. (1975) *Ordnance Survey Maps: A Descriptive Manual*. Southampton: Ordnance Survey.

Haslett, J., Bradley, R., Craig, P., Unwin, A. and Wills, G. (1991) 'Dynamic Graphics for Exploring Spatial Data with Applications to Locating Global and Local Anomalies', *The American Statistician*, 45: 234–42.

Haslett, J., Wills, G. and Unwin, A. (1990) 'SPIDER, an Interactive Statistical Tool for the Analysis of Spatially Distributed Data', *International Journal of Geographical Information Systems*, 4: 285–96.

Hastie, T. and Tibshirani, R. (1990) *Generalized Additive Models.* London: Chapman and Hall.

Hauser, R.M. (1970) 'Context and Consex: A Cautionary Tale', *American Journal of Sociology,* 75: 645–64.

Haynes, K.E. and Fotheringham, A.S. (1984) *Gravity and Spatial Interaction Models,* Vol. 2, Sage Series in Scientific Geography. Beverly Hills, CA: Sage.

Haynes, K.E. and Fotheringham, A.S. (1990) 'The Impact of Space on the Application of Discrete Choice Models', *Review of Regional Studies,* 20: 39–49.

Hepple, L. (1974) 'The Impact of Stochastic Process Theory upon Spatial Analysis', *Progress in Human Geography,* 6: 89–142.

Hepple, L. (1998) 'Context, Social Construction and Statistics: Regression, Social Science and Human Geography', *Environment and Planning A,* 30: 225–34.

Hertz, J., Krogh, A. and Palmer, J.A. (1991) *Introduction to the Theory of Neural Computation.* Reading, MA: Addison-Wesley.

Heuvelink, G.M.B., Burrough, P.A. and Stein, A. (1989) 'Propagation of Errors in Spatial Modelling with GIS', *International Journal of Geographical Information Systems,* 3: 303–22.

Heywood, I., Cornelius, S. and Carver, S.J. (1998) *An Introduction to Geographical Information Systems.* Harlow: Addison Wesley Longman.

Hinckley, D. (1988) 'Bootstrap Methods', *Journal of the Royal Statistical Society,* B, 50: 321–37.

Hirtle, S.C. and Jonides, J. (1985) 'Evidence of Hierarchies in Cognitive Maps', *Memory and Cognition,* 13: 208–17.

Hoffman, P., Grinstein, G., Marx, K., Grosse, I. and Stanley, E. (1997) 'DNA Visual and Analytic Data Mining', *I.E.E.E. Visualization '97 Proceedings,* 437–41, Phoenix, AZ.

Holt, D., Steel, D.G., Tranmer, M. and Wrigley, N. (1996) 'Aggregation and Ecological Effects in Geographically Based Data', *Geographical Analysis,* 28: 244–61.

Hordijk, L. (1974) 'Spatial Correlation in the Disturbances of a Linear Interregional Model', *Regional and Urban Economics,* 4: 117–40.

Huber, J.J., Payne, W. and Pluto, C. (1982) 'Adding a Symmetrically Dominated Alternative: Violations of Regularity and the Similarity Hypothesis', *Journal of Consumer Research,* 9: 90–8.

Huber, P.J. (1985) 'Projection Pursuit' (with discussion), *The Annals of Statistics,* 13: 435–525.

Huff, D.L. (1959) 'Geographical Aspects of Consumer Behavior', *University of Washington Business Review,* 18: 27–37.

Huff, D.L. (1963) 'A Probabilistic Analysis of Consumer Behavior', *Papers and Proceedings of the Regional Science Association,* 7: 81–90.

Huxhold, W. (1991) *An Introduction to Urban Geographic Information Systems.* Oxford: Oxford University Press.

Ihaka, R. and Gentleman, R. (1996) 'R: A Language for Data Analysis and Graphics', *Journal of Computational and Graphical Statistics,* 5: 299–314.

Inselberg, A. (1985) 'The Plane with Parallel Coordinates', *The Visual Computer,* 1: 69–91.

Inselberg, A. (1988) 'Visual Data Mining with Parallel Co-ordinates', *Computational Statistics,* 13: 47–63.

Isaaks, E. and Srivastava, R. (1988) *An Introduction to Applied Geostatistics.* Oxford: Oxford University Press.

Jaynes, E.T. (1957) 'Information Theory and Statistical Mechanics', *Physical Review,* 106: 620–30.

Johnston, R.J. (1994) 'Spatial Analysis', in R.J. Johnston, D. Gregory and D.M. Smith (eds), *The Dictionary of Human Geography.* Oxford: Blackwell. p. 577.

Johnston, R.J. (1997) 'W(h)ither Spatial Science and Spatial Analysis?', *Futures*, 29: 323–36.

Johnston, R.J., Pattie, C.J. and Allsop, J.G. (1988) *A Nation Dividing?*. London: Longman.

Jones, J.P. and Casetti, E. (1992) *Applications of the Expansion Method*. London: Routledge.

Jones, J.P. and Hanham, R.Q. (1995) 'Contingency, Realism, and the Expansion Method', *Geographical Analysis*, 27: 185–207.

Jones, K. (1991a) 'Specifying and Estimating Multilevel Models for Geographical Research', *Transactions of The Institute of British Geographers*, 16: 148–59.

Jones, K. (1991b) *Multilevel Models for Geographical Research*, Concepts and Techniques in Modern Geography, 54. Norwich: Environmental Publications.

Jones, K. (1997) 'Multilevel Approaches to Modelling Contextuality: From Nuisance to Substance in the Analysis of Voting Behaviour', in G.P. Westert and R.N. Verhoeff (eds), *Places and People: Multilevel Modelling in Geographical Research,* Nederlandse Geografische Studies 227, University of Utrecht. pp. 19–40.

Jones, K. and Bullen, N.J. (1993) 'A Multilevel Analysis of the Variations in Domestic Property Prices: Southern England', *Urban Studies*, 30: 1409–26.

Jones, K., Gould, M.I. and Watt, R. (1996) 'Multiple Contexts as Cross-Classified Models: the Labour Vote in the British General Election of 1992', Mimeo, Department of Geography, University of Portsmouth.

Jones, M. and Sibson, R. (1987) 'What is Projection Pursuit?' (with discussion), *Journal of the Royal Statistical Society*, 150: 1–36.

Jong, de T. and Ottens, H. (1997) 'GIS Functionality for Multi-level Research' in G.P. Westert and R.N. Verhoeff (eds), *Places and People: Multilevel Modelling in Geographical Research*, Nederlandse Geografische Studies 227, University of Utrecht. pp. 44–54.

Kelsall, J.E. and Diggle, P.J. (1995) 'Non-parametric Estimation of Spatial Variation in Relative Risk', *Statistics in Medicine*, 14: 2335–42.

Kendall, S.M. and Ord, J.K. (1973) *Time Series*. Sevenoaks, Kent: Edward Arnold.

Kerhis, E. (1989) *Interfacing Arc/Info and GLIM: A Progress Report*, Research Report 5, NorthWest Regional Research Laboratory, University of Lancaster.

King, G. (1997) *A Solution to the Ecological Inference Problem: Reconstructing Individual Behavior from Aggregate Data*. Princeton, NJ: Princeton University Press.

King, L.J. (1961) 'A Multivariate Analysis of the Spacing of Urban Settlements in the United States', *Annals of the Association of American Geographers*, 51: 222–33.

King, L.J. (1969) *Statistical Analysis in Geography*. Englewood Cliffs, NJ: Prentice Hall.

Knorrheld, L. and Besag, J. (1998) 'Modelling Risk from a Disease in Time and Space', *Statistics in Medicine*, 17: 2045–60.

Knox, E.G. (1964) 'Epidemiology of Childhood Leukaemia in Northumberland and Durham', *British Journal of Preventative and Social Medicine*, 18: 17–24.

Kohonen, T. (1989) *Self Organization and Associative Memory*. Berlin: Springer-Verlag.

Kremenec, A.J. and Esparza, A. (1993) 'Modeling Interaction in a System of Markets', *Geographical Analysis*, 25: 354–68.

Krige, D. (1966) 'Two-dimensional Weighted Moving Average Surfaces for Ore Evaluation', *Journal of South African Institute of Mining and Metallurgy*, 66: 13–38.

Krugman, P. (1996) 'Urban Concentration: The Role of Increasing Returns and Transport Costs', *International Regional Science Review*, 19: 5–30.

Kruskal, J.B. (1969) 'Toward a Practical Method which Helps Uncover the Structure of a Set of Observations by Finding the Linear Transformation which Optimizes a New "Index of Condensation"', in R. Milton and J.A. Nelder (eds), *Statistical Computation*. New York: Academic Press. pp. 427–40.

Laurini, R. and Thompson, D. (1992) *Fundamentals of Spatial Information Systems*. London: Academic Press.

Lee, P.M. (1997) *Bayesian Statistics: An Introduction.* London: Arnold.

Leeuw, J. de (1994) 'Statistics and the Sciences', Unpublished manuscript, UCLA Statistics Program.

Lillesand, T.M. and Kiefer, R.W. (1994) *Remote Sensing and Image Interpretation.* New York: Wiley.

Lindsay, P.H. and Norman, D.A. (1972) *Human Information Processing.* New York: Academic Press.

Linneman, H.V. (1966) *An Econometric Study of International Trade Flows.* Amsterdam: North-Holland.

Lo, L. (1991) 'Substitutability, Spatial Structure, and Spatial Interaction', *Geographical Analysis*, 23: 132–46.

Loh, W.-Y. and Wu, C.F.J. (1987) 'Comment on Efron', *Journal of the American Statistical Association*, 82: 188–90.

Longley, P. and Batty, M. (1996) *Spatial Analysis: Modelling in a GIS Environment.* Cambridge: GeoInformation International.

Longley, P., Brooks, S.M., McDonnell, R. and Macmillan, B. (eds) (1998) *Geocomputation: A Primer.* Chichester: Wiley.

MacEachren, A.M., Brewer, C.A. and Pickle, L.W. (1998) 'Visualizing Georeferenced Data: Representing Reliability of Health Statistics', *Environment and Planning A*, 30: 1547–61.

Maddala, G.S. (1977) *Econometrics.* New York: McGraw-Hill.

Majure, J. and Cressie, N. (1997) 'Dynamic Graphics for Exploring Spatial Dependence in Multivariate Spatial Data', *Geographical Systems*, 4: 131–58.

Maling, D.H. (1993) *Coordinate Systems and Map Projections.* 2nd edn. Oxford: Pergamon.

Mallows, C. (1973) 'Some Comments on C_p', *Technometrics*, 15: 661–7.

Mantel, N. (1967) 'The Detection of Disease Clustering and a Generalised Regression Approach', *Cancer Research*, 27: 209–20.

Mardia, K., Kent, J. and Bibby, J.M. (1979) *Multivariate Analysis.* London: Academic Press.

Marshall, R.J. (1991) 'A Review of Methods for the Statistical Analysis of Spatial Patterns of Disease', *Journal of the Royal Statistical Society*, A, 154: 421–41.

Martin, D. (1996) *Geographic Information Systems: Socio-economic Applications.* London: Routledge.

McCarty, H. (1956) 'Use of Certain Statistical Procedures in Geographical Analysis', *Annals of the Association of American Geographers*, 46: 263.

McCulloch, W. and Pitts, W. (1943) 'A Logical Calculus of Ideas Immanent in Neural Activity', *Bulletin of Mathematical Biophysics*, 5: 115–33.

McFadden, D. (1974) 'Conditional Logit Analysis of Qualitative Choice Behavior', in P. Zarembka (ed.), *Frontiers in Econometrics.* New York: Academic Press. pp. 105–42.

McFadden, D. (1978) 'Modelling the Choice of Residential Location', in A. Karlquist, L. Lundquist, F. Snickars and J.W. Weibull (eds), *Spatial Interaction Theory and Planning Models.* Amsterdam: North-Holland. pp. 75–96.

McFadden, D. (1980) 'Econometric Models for Probabilistic Choice Among Products', *Journal of Business*, 53: 513–29.

McNamara, T.P. (1986) 'Mental Representations of Spatial Relations', *Cognitive Psychology*, 18: 87–121.

McNamara, T.P. (1992) 'Spatial Representation', *Geoforum*, 23: 139–50.

Menzel, H. (1950) 'Comment on Robinson's "Ecological Correlations and the Behaviour of Individuals"', *American Sociological Review*, 15: 674.

Metropolis, N. and Ulam, S. (1949) 'The Monte-Carlo Method', *Journal of the American Statistical Association*, 44: 335–41.

Metropolis, N. Rosenbluth, A., Rosenbluth, A.W. and Teller, A.H. (1949) 'Equation of State Calculations by Fast Computing Machines', *Journal of Chemical Physics*, 21: 1087–92.

Meyer, R.J. (1979) 'Theory of Destination Choice-Set Formation under Informational Constraints', *Transportation Research Record*, 750: 6–12.

Meyer, R.J. and Eagle, T.C. (1982) 'Context-Induced Parameter Instability in a Disaggregate-Stochastic Model of Store Choice', *Journal of Marketing Research*, 19: 62–71.

Miller, H.J. (1999) 'Measuring Space-Time Accessibility Benefits with Transportation Networks: Basic Theory and Computational Procedures', *Geographical Analysis*, 31: 187–212.

Miyares, I.M. and McGlade, M.S. (1994) 'Specialization in "Jobs in Geography" 1980–1993', *The Professional Geographer*, 46: 170–7.

Moellering, H. and Tobler, W.R. (1972) 'Geographical Variances', *Geographical Analysis*, 4: 34–50.

Molho, I. (1995) 'Spatial Autocorrelation in British Unemployment', *Journal of Regional Science*, 35: 641–58.

Mooney, C.Z. and Duvall, R.D. (1993) *Bootstrapping: A Nonparametric Approach to Statistical Inference*. London: Sage.

Morrill, R.L. (1984) 'Recollections of the Quantitative Revolution's Early Years: The University of Washington 1955–65', in M. Billinge, D. Gregory and R. Martin (eds), *Recollections of a Revolution*. New York: St. Martin's Press. pp. 57–72.

Nelder, J. (1971) 'Discussion on Papers by Wynn and Bloomfield, and O'Neill and Wetherill', *Journal of the Royal Statistical Society*, B, 33: 244–6.

Nester, M. (1996) 'An Applied Statistician's Creed', *Applied Statistics*, 45: 401–10.

Newell, A. and Simon, H.A. (1972) *Human Problem Solving*. Englewood Cliffs, NJ: Prentice Hall.

Niedercorn, J.H. and Bechdolt Jr, V.B. (1969) 'Economic Derivation of the "Gravity Law" of Spatial Interaction', *Journal of Regional Science*, 9: 273–82.

Norman, D.A. and Bobrow, D.G. (1975) 'On Data-Limited and Resource-Limited Processes', *Cognitive Psychology*, 7: 44–64.

Odland, J. (1988) *Spatial Autocorrelation*, Scientific Geography Series, 9. Newbury Park, CA: Sage.

O'Loughlin, J., Ward, M.D., Lofdahl, C.L., Cohen, J.S., Brown, D.S., Reilly, D., Gleditsch, K.S. and Shin, M. (1998) 'The Diffusion of Democracy 1946–1994', *Annals of the Association of American Geographers*, 88: 545–74.

O'Neill, R. and Wetherill, G. (1971) 'The Present State of Multiple Comparison Methods', *Journal of the Royal Statistical Society*, B, 33: 218–41.

Openshaw, S. (1984) *The Modifiable Areal Unit Problem*, CATMOG 38. Norwich: GeoAbstracts.

Openshaw, S. (1993) 'Exploratory Space-Time-Attribute Pattern Analysers', in A.S. Fotheringham and P.A. Rogerson (eds), *Spatial Analysis and GIS*. London: Taylor and Francis. pp. 147–63.

Openshaw, S. and Abrahart, R.J. (1996) 'GeoComputation', Abstracted in *Proceedings Geocomputation '96, 1st International Conference on Geocomputation, University of Leeds, 17–19 September*. pp. 665–6.

Openshaw, S. and Openshaw, C. (1997) *Artificial Intelligence in Geography*. Chichester: Wiley.

Openshaw, S. and Rao, L. (1995) 'Algorithms for Re-engineering 1991 Census Geography', *Environment and Planning A*, 27: 425–46.

Openshaw, S. and Taylor, P.J. (1979) 'A Million or so Correlation Coefficients: Three Experiments on the Modifiable Areal Unit Problem', in N. Wrigley (ed.), *Statistical Applications in the Spatial Sciences*. London: Pion. pp. 127–44.

Openshaw, S., Abrahart, R.J. and Harris, T.E. (eds) (1999) *Geocomputation*. London: Taylor and Francis. at press.

Openshaw, S., Brunsdon, C. and Charlton, M.E. (1991a) 'A Spatial Analysis Toolkit for GIS', *Proceedings EGIS '91*. Brussels: EGIS Foundation, 2: 788–96.

Openshaw, S., Charlton, M.E. and Carver, S.J. (1991b) 'Error Propagation: A Monte Carlo Simulation', in I. Masser and M. Blakemore (eds), *Handling Geographic Information*. Harlow: Longman. pp. 78–101.

Openshaw, S., Charlton, M.E., Wymer, C. and Craft, A.W. (1987) 'A Mark I Geographical Analysis Machine for the Automated Analysis of Point Data Sets', *International Journal of Geographical Information Systems*, 1: 359–77.

Openshaw, S., Craft, A. and Charlton, M.E. (1988) 'Searching for Leukaemia Clusters using a Geographical Analysis Machine', *Papers of the Regional Science Association*, 64: 95–106.

Ord, J.K. (1975) 'Estimation Methods for Models of Spatial Interaction', *Journal of the American Statistical Association*, 70: 120–6.

Ord, J.K. and Getis, A. (1995) 'Local Spatial Autocorrelation Statistics: Distributional Issues and an Application', *Geographical Analysis*, 27: 286–306.

Ordnance Survey (no date) ED–LINE, *http://www.ordsvy.gov.uk/products/computer/ed–line/index.htm*, accessed 21 March 1999.

O'Rourke, J. (1998) *Computational Geometry in C.* Cambridge: Cambridge University Press.

Parzen, E. (1962) 'On the Estimation of a Probability Density Function and Model', *Annals of Mathematical Statistics*, 33: 1065–76.

Pellegrini, P.A. and Fotheringham, A.S. (1999) 'Intermetropolitan Migration and Hierarchical Destination Choice: A Disaggregate Analysis from the US PUMS', *Environment and Planning A*, 31: 1093–118.

Peters, A. (1989) *Peters' Atlas of the World.* London: Longman.

Pipkin, J.S. (1978) 'Fuzzy Sets and Spatial Choice', *Annals of the Association of American Geographers*, 68: 196–204.

Pooler, J. (1987) 'Measuring Geographical Accessibility: A Review of Current Approaches and Problems in the Use of Population Potentials', *Geoforum*, 18: 269–89.

Powe, N.A., Garrod, G.D., Brunsdon, C. and Willis, K.G. (1997) 'Using a Geographic Information System to Estimate an Hedonic Price Model of the Benefits of Woodland Access', *Forestry*, 70: 139–49.

Raper, J., Rhind, D. and Shepherd, J. (1992) *Postcodes: The New Geography.* Harlow: Longman.

Rasbash, J. and Woodhouse, G. (1995) *Mln Command Reference Version 1.0.* Multilevel Models Project, Institute of Education, University of London.

Ravenstein, E.G. (1885) 'The Laws of Migration', *Journal of the Royal Statistical Society*, 48: 167–235.

Reed, R. (1993) 'Pruning Algorithms – a survey', *IEEE Transactions on Neural Networks*, 4: 740–7.

Rees, P. (1995) 'Putting the Census on the Researcher's Desk', in S. Openshaw (ed.) *The Census Users' Handbook.* Cambridge: GeoInformation International. pp. 27–81.

Rhind, D., Goodchild, M. and Maguire, D. (1991) *Geographic Information Systems.* London: Longman.

Ripley, B. (1981) *Spatial Statistics.* Chichester: Wiley.

Robinson, A.H. (1956) 'The Necessity of Weighting Values in Correlation Analysis of Areal Data', *Annals of the Association of American Geographers*, 46: 233–6.

Robinson, G., Peterson, J.A. and Anderson, P.A. (1971) 'Trend Surface Analysis of Crime Attitudes in Scotland', *Scottish Geographical Magazine*, 87: 142–6.

Robinson, G.M. (1998) *Methods and Techniques in Human Geography.* Chichester: Wiley.

Robinson, W.R. (1950) 'Ecological Correlation and the Behaviour of Individuals', *American Sociological Review*, 15: 351–7.

Rose, J.K. (1936) 'Corn Yield and Climate in the Corn Belt', *The Geographical Review*, 26: 88–102.

Rowlingson, B. and Diggle, P. (1993) 'Splancs: Spatial Point Pattern Analysis Code in S-Plus', *Computers and Geosciences*, 19: 627–55.

Sampson, P.D. and Guttorp, P. (1992) 'Nonparametric Estimation of Nonstationary Spatial Covariance Structure', *Journal of the American Statistical Association*, 87: 108–19.

Savage, I. (1957) 'Nonparametric Statistics', *Journal of the American Statistical Association*, 52: 331–44.

Sayer, A. (1976) 'A Critique of Urban Modelling', *Progress in Planning*, 6: 187–254.

Sayer, A. (1992) *Method in Social Science*. London: Routledge.

Schettino, A. (1999) 'Polygon Intersections in Spherical Topology: Application to Plate Tectonics', *Computers and Geosciences*, 25: 61–9.

Scott, D. (1992) *Multivariate Density Estimation: Theory, Practice and Visualisation*. New York: Wiley.

Sedgwick, R. (1990) *Algorithms in C*. Reading, MA: Addison-Wesley.

Seimiatycki, J., Brubaker, G. and Geser, A. (1980) 'Space-Time Clustering of Burkitt's Lymphoma in East Africa: Analysis of Recent Data and a New Look at Old Data', *International Journal of Cancer*, 25: 197–203.

Sen, A. and Smith, T.E. (1995) *Gravity Models of Spatial Interaction Behavior*. Berlin: Springer-Verlag.

Shannon, C.F. (1948) 'A Mathematical Theory of Communication', *Bell System Technical Journal*, 27: 379–423 and 623–56.

Silverman, B.W. (1986) *Density Estimation for Statistics and Data Analysis*. London: Chapman and Hall.

Simon, H.A. (1969) *The Science of the Artificial*. Cambridge, MA: MIT Press.

Smit, L. (1997) 'Changing Commuter Distances in the Netherlands: A Macro-Micro Perspective', in G.P. Westert and R.N. Verhoeff (eds), *Places and People: Multilevel Modelling in Geographical Research*, Nederlandse Geografische Studies 227, University of Utrecht. pp. 86–99.

Sokal, R.R., Oden, N.L. and Thomson, B.A. (1998) 'Local Spatial Autocorrelation in a Biological Model', *Geographical Analysis*, 30: 331–54.

Steel, D.G. and Holt, D. (1996) 'Rules for Random Aggregation', *Environment and Planning A*, 28: 957–78.

Stephan, F.F. (1934) 'Sampling Errors and Interpretations of Social Data Ordered in Time and Space', *Journal of the American Statistical Association*, 29: 165–6.

Stevens, A. and Coupe, P. (1978) 'Distortions in Judged Spatial Relations', *Cognitive Psychology*, 10: 422–37.

Stevens, S.S. (1957) 'On the Psychophysical Law', *Psychological Review*, 64: 153–81.

Stevens, S.S. (1975) *Psychophysics: Introduction to its Perceptual, Neural and Social Prospects*. New York: Wiley.

Stone, R.A. (1988) 'Investigations of Excess Environmental Risks around Putative Sources: Statistical Problems and a Proposed Test', *Statistical Methods*, 7: 649–60.

Strahler, A.N. (1952) 'Hypsometric (area-altitude) Analysis of Erosional Topography', *Bulletin of the Geological Society of America*, 63: 1117–42.

Swayne, D., Cook, D. and Buja, A. (1991) 'XGobi: Interactive Dynamic Graphics in the X Window System with a Link to S', *American Statistical Association Proceedings of the Section on Statistical Graphics*, pp. 1–8.

Taylor, P.J. and Johnston, R.J. (1995) 'Geographic Information Systems and Geography', in J. Pickles (ed.), *Ground Truth*. London: Guilford. pp. 51–67.

Terrell, G. (1990) 'The Maximal Smoothing Principal in Density Estimation', *Journal of the American Statistical Association*, 85: 470–7.

Terrell, G. and Scott, D. (1985) 'Oversmoothed Nonparametric Density Estimations', *Journal of the American Statistical Association*, 80: 209–14.

Thill, J.-C. (1992) 'Choice Set Formation for Destination Choice Modelling', *Progress in Human Geography*, 16: 361–82.

Thomas, E.N. and Anderson, D.L. (1965) 'Additional Comments on Weighting Values in Correlation Analysis of Areal Data', *Annals of the Association of American Geographers*, 55: 492–505.

Tiefelsdorf, M. (1998) 'Some Practical Applications of Moran's I's Exact Conditional Distribution', *Papers of the Regional Science Association*, 77: 101–29.

Tiefelsdorf, M. and Boots, B. (1997) 'A Note on the Extremities of Local Moran's I's and their Impact on Global Moran's I', *Geographical Analysis*, 29: 248–57.

Tiefelsdorf, M., Fotheringham, A.S. and Boots, B. (1998) 'Exploratory Identification of Global and Local Heterogeneities in Disease Mapping', Presented at the Association of American Geographers' Annual Meeting, Boston.

Tierney, L. (1990) *LISP-STAT: An Object-oriented Environment for Statistical Computing and Dynamic Graphics.* New York: Wiley.

Tinkler, K.J. (1971) 'Statistical Analysis of Tectonic Patterns in Areal Volcanism: The Bunyaraguru Volcanic Field in Western Uganda', *Mathematical Geology*, 3: 335–55.

Tobler, W.R. (1963) 'Geographic Area and Map Projections', *Geographical Review*, 53: 59–78.

Tobler, W.R. (1967) 'Automated Cartograms', Mimeo.

Tobler, W.R. (1970) 'A Computer Movie Simulating Urban Growth in the Detroit Region', *Economic Geography*, 46: 234–40.

Tobler, W.R. (1973a) 'Choropleth Maps without Class Intervals?', *Geographical Analysis*, 3: 262–5.

Tobler, W.R. (1973b) 'A Continuous Transformation useful for Districting', *Annals of the New York Academy of Sciences*, 219: 215–20.

Tobler, W.R. (1979) 'Smooth Pycnophylactic Interpolation for Geographical Regions', *Journal of the American Statistical Association*, 74: 121–7.

Tobler, W.R. (1991) 'Frame Independent Spatial Analysis', in M. Goodchild and S. Gopal (eds), *The Accuracy of Spatial Databases.* London: Taylor and Francis. pp. 115–22.

Tomlin, C.D. (1990) *Geographic Information Systems and Cartographic Modeling.* Eaglewood Cliffs, NJ: Prentice Hall.

Tranmer, M. and Steel, D.G. (1998) 'Using Census Data to Investigate the Causes of the Ecological Fallacy', *Environment and Planning A*, 30: 817–31.

Tribus, M. (1969) *Rational Descriptions, Decisions and Designs.* New York: Pergamon.

Tufte, E. (1983) *The Visual Display of Quantitative Information.* Cheshire, CT: Graphics Press.

Tukey, J. (1977) *Exploratory Data Analysis.* Reading, MA: Addison-Wesley.

Unwin, A. and Unwin, D. (1998) 'Exploratory Spatial Data Analysis with Local Statistics', *The Statistician*, 47: 415–23.

Unwin, D. (1981) *Introductory Spatial Analysis.* London: Methuen.

Unwin, D. (1996) 'GIS, Spatial Analysis and Spatial Statistics', *Progress in Human Geography*, 20: 540–51.

Venables, W.N. and Ripley, B.D. (1997) *Modern Applied Statistics with S-Plus*, 2nd edn. New York: Springer-Verlag.

Ver Hoef, J.M. and Cressie, N. (1993) 'Multivariable Spatial Prediction', *Mathematical Geology*, 25: 219–40.

Verheij, R.A. (1997) 'Physiotherapy Utilization: Does Place Matter?', in G.P. Westert and

R.N. Verhoeff (eds), *Places and People: Multilevel Modelling in Geographical Research*, Nederlandse Geografische Studies 227, University of Utrecht. pp. 74–85.

Vincent, P. and Gatrell, A. (1991) 'The Spatial Distribution of Radon Gas in Lancashire (UK): A Kriging Study', *Proceedings, Second European Conference on Geographical Information Systems*. Utrecht: EGIS Foundation. pp. 1179–86.

Walker, R. and Calzonetti, F. (1989) 'Searching for New Manufacturing Plant Locations: A Study of Location Decisions in Central Appalachia', *Regional Studies*, 24: 15–30.

Watt, R. (1991) *Understanding Vision*. London: Academic Press.

Webber, M.J. (1975) 'Entropy-Maximising Location Models for Nonindependent Events', *Environment and Planning A*, 7: 99–108.

Webber, M.J. (1977) 'Pedagogy Again: What is Entropy?', *Annals of the Association of American Geographers*, 67: 254–66.

Weber, A. (1909) *Theory of the Location of Industries*. Chicago: University of Chicago Press.

Weibull, J.W. (1976) 'An Axiomatic Approach to the Measurement of Accessibility', *Regional Science and Urban Economics*, 6: 357–79.

Weibull, J.W. (1980) 'On the Numerical Measurement of Accessibility', *Environment and Planning A*, 12: 53–67.

Westert, G.P. and Verhoeff, R.N. (eds) (1997) *Places and People: Multilevel Modelling in Geographical Research*, Nederlandse Geografische Studies 227, University of Utrecht.

White, H. (1988) 'Economic Prediction using Neural Networks: The case of IBM Daily Stock Returns', in *Proceedings of the IEEE International Conference on Neural Networks, San Diego*, Vol. II, pp. 451–9.

Wilson, A.G. (1967) 'Statistical Theory of Spatial Trip Distribution Models', *Transportation Research*, 1: 253–69.

Wilson, A.G. (1974) *Urban and Regional Models in Geography and Planning*. London: Wiley.

Wilson, A.G. (1975) 'Some New Forms of Spatial Interaction Models: A Review', *Transportation Research*, 9: 167–79.

Wilson, A.G. (1984) 'One Man's Quantitative Geography: Frameworks, Evaluations, Uses and Prospects', in M. Billinge, D. Gregory and R. Martin (eds), *Recollections of a Revolution: Geography as a Spatial Science*. New York: St. Martin's Press. pp. 200–26.

Worboys, M. (1995) *GIS: A Computing Perspective*. London: Taylor and Francis.

Xia, F.F. and Fotheringham, A.S. (1993) 'Exploratory Spatial Data Analysis with GIS: The Development of the ESDA Module under Arc/Info', *GIS/LIS '93 Proceedings*, 2: 801–10.

Yule, G.U. and Kendall, M.G. (1950) *An Introduction to Statistics*. New York: Hafner.

Zadeh, L. (1965) 'Fuzzy Sets', *Information and Control*, 8: 338–53.

Zipf, G.K. (1949) *Human Behavior and the Principle of Least Effort*. Reading, MA: Addison-Wesley.

Index

accessibility
    discussion of definition 243
    role in spatial interaction
        modelling 231–2
autoregressive models 166–71

bandwidth
    calibration of 112, 148–9
    in kernel intensity estimates 147–9
    and smoothing 46
Bayes theorem 193–5
Bayesian inference 193–8, 240
box plots
    of burglary data 138
    description of 68–70
    of health data 127
bubble plot 134
buffering 34–5

cartograms 20, 75
classical inference 198–201
cluster analysis 188–90
competing destinations model
    225–34
computational power 7, 13
confidence intervals
    in classical inference 200–1
    of Moran's I 205–6, 208
contiguity 41–2
coordinate systems
    latitude and longitude 18–20
    spherical 241–2
cross–area aggregation 59–60
cross-validation
    and geographically weighted
        regression 113

and semi-parametric smoothing
    180

data mining 187–93
digital elevation model 43
digitizing 16–17
discrete choice modelling 222–5
disease clusters 9
distance 20–1
distance–decay parameters
    with misspecification bias 226–7
    without misspecification bias
        233–4

EM algorithm 59–60
empiricism 7
error modelling 58–9
expansion method
    example of 117–18
    spatial version defined 106–7
experimental distributions 204–9
exploratory data analysis
    definition of 8
    examples of 65–92
    and GIS 54–5
    pre-modelling and post-modelling
        10
    and visualization 185–7

fields
    approximation of 21–2
    interpolation and 42–5
    measurement of 18
first order intensity analysis 144–9

five number summary
  description of 68–9
  example of 138
frequency Polygons 70–1

G statistic 99–101
genealized cross validation score
    180–1
geocomputation 12
geographical analysis machine 96–7
geographically weighted regression
  description of 107–28
  extensions to other statistics 239
geometric intersection 36–8
GIS
  and advanced types of spatial
      analysis 49–58
  linking 61–3
  and local forms of spatial analysis
      94
  problems with 58–60
  and quantitative methods 10,
      30–2
  and simple types of spatial analysis
      32–49
  and spatial data 15–16
GPS 17
grand tour 90–1

hedonic price models 51–4
histograms
  bin widths 71
  general discussion 70

IIA property 225
independence 26
index of dispersion 145
informal inference 185
interpolation 42–5

K functions
  comparison of 157–9

general discussion with examples
    149–54
kernel density estimation
  and bandwidths 71–2
  comparing kernel densities 155–7
  general concept 45–6
  and geographically weighted
      regression 108–9
  and point patterns 146–9
  spatially adaptive kernels 111–2
Kriging 171–8

linked plots
  projection pursuit and maps 90
  RADVIZ and maps 86
  scatterplots and maps 77–9
local analysis
  basic idea of 11
  an empirical example 114–28
  general issues in 93–6
  with geographically weighted
      regression 107–14
  local point pattern analysis 96–7
  multivariate 102–7
  in spatial interaction modelling
      128–9, 226–7, 233–4
  univariate 97–102

maps 72–5
maximum likelihood estimators 200
mean centre 136
mean integrated squared error
    148–9
mean nearest neighbour distance
    137
model building and testing 210–11
modifiable areal unit problem 28
  overview and prospects 235–8
  in point patterns 186–7
Moran scatterplot 99
Moran's *I*
  definition and theoretical variance
      202–4
  an empirical of 206–9

experimental distribution for
204–6
local version of 101–2
moving average models 169–71
multilevel modelling 103–6
multiple significance testing 210

networks 22–3, 46
neural networks 190–3

parallel coordinates plots 80–2
point-in-polygon operations 24–5,
32
point pattern analysis
comparisons of point patterns
154–60
first order processes 144–9
general concepts in 130–2
local versions of 96–7
modelling 138–44
non-graphical approaches 136–8
second order processes 149–54
and visualization 132–6
positivism 5
postcodes 16, 42
projection pursuit 86–91
projections 19–20
proportional symbol plot 134

quantitative revolution 234
querying 49

RADVIZ 82–6
random dot maps 74–6
randomness 8
raster model 23–4
remotely sensed images 17

scatterplot matrix 75–8
second order intensity analysis
149–54
semi–parametric smoothing 178–82

semi-variogram 174–6
slicing and dynamic plots 90
social physics 215–17
space-time models 243–4
spatial association statistics 99–101
spatial autocorrelation
definition of 12–13
effects on the distribution of the
sample mean 27–8
local measure of 101–2
and Moran's *I* scatterplot 99
and spatially autoregressive
models 167–71
and statistical inference 201–9
spatial autoregressive models 167–9
spatial choice 227–34
spatial cognition 13, 228–33
spatial data sets 8
spatial information processing
225–34
spatial interaction models
and discrete choice modelling
222–5
local versions of 128–9
phases in the development of
213–35
and social physics 215–17
and spatial information processing
225–34
and statistical mechanics 217–22
spatial moving average models
169–71
spatial non-stationarity
challenges in 238–40
and local models 93–129
spatial objects 17–18, 21
spatial regression models 11–12,
162–71
standard distance 136
statistical inference for spatial data
184–212
statistical mechanics 217–21
stem and leaf plots 66–8

Theissen polygons 38–41
theory 7

uncertainty 6
unclassed choropleth maps 73, 75

variogram cloud plot 97–8
vector model 22–3
visualization
  current status 10–11
  and inference 185–7

of model outputs 55–8
overview 65–6
and surfaces 43–4
techniques for bivariate data 75–9
techniques for multivariate data
  80–91
techniques for univariate data
  66–75